P9-DUK-302

Lecture Notes in Physics

Springer
Berlin
Heidelberg
New York
Barcelona
Budapest
Hong Kong
London
Milan
Paris
Santa Clara
Singapore
Tokyo

The Editorial Policy for Proceedings

The series Lecture Notes in Physics reports new developments in physical research and teaching – quickly, informally, and at a high level. The proceedings to be considered for publication in this series should be limited to only a few areas of research, and these should be closely related to each other. The contributions should be of a high standard and should avoid lengthy redraftings of papers already published or about to be published elsewhere. As a whole, the proceedings should aim for a balanced presentation of the theme of the conference including a description of the techniques used and enough motivation for a broad readership. It should not be assumed that the published proceedings must reflect the conference in its entirety. (A listing or abstracts of papers presented at the meeting but not included in the proceedings could be added as an appendix.)
When applying for publication in the series Lecture Notes in Physics the volume's editor(s) should submit sufficient material to enable the series editors and their referees to make a fairly accurate evaluation (e.g. a complete list of speakers and titles of papers to be presented and abstracts). If, based on this information, the proceedings are (tentatively) accepted, the volume's editor(s), whose name(s) will appear on the title pages, should select the papers suitable for publication and have them refereed (as for a journal) when appropriate. As a rule discussions will not be accepted. The series editors and Springer-Verlag will normally not interfere with the detailed editing except in fairly obvious cases or on technical matters.
Final acceptance is expressed by the series editor in charge, in consultation with Springer-Verlag only after receiving the complete manuscript. It might help to send a copy of the authors' manuscripts in advance to the editor in charge to discuss possible revisions with him. As a general rule, the series editor will confirm his tentative acceptance if the final manuscript corresponds to the original concept discussed, if the quality of the contribution meets the requirements of the series, and if the final size of the manuscript does not greatly exceed the number of pages originally agreed upon. The manuscript should be forwarded to Springer-Verlag shortly after the meeting. In cases of extreme delay (more than six months after the conference) the series editors will check once more the timeliness of the papers. Therefore, the volume's editor(s) should establish strict deadlines, or collect the articles during the conference and have them revised on the spot. If a delay is unavoidable, one should encourage the authors to update their contributions if appropriate. The editors of proceedings are strongly advised to inform contributors about these points at an early stage.
The final manuscript should contain a table of contents and an informative introduction accessible also to readers not particularly familiar with the topic of the conference. The contributions should be in English. The volume's editor(s) should check the contributions for the correct use of language. At Springer-Verlag only the prefaces will be checked by a copy-editor for language and style. Grave linguistic or technical shortcomings may lead to the rejection of contributions by the series editors. A conference report should not exceed a total of 500 pages. Keeping the size within this bound should be achieved by a stricter selection of articles and not by imposing an upper limit to the length of the individual papers. Editors receive jointly 30 complimentary copies of their book. They are entitled to purchase further copies of their book at a reduced rate. As a rule no reprints of individual contributions can be supplied. No royalty is paid on Lecture Notes in Physics volumes. Commitment to publish is made by letter of interest rather than by signing a formal contract. Springer-Verlag secures the copyright for each volume.

The Production Process

The books are hardbound, and the publisher will select quality paper appropriate to the needs of the author(s). Publication time is about ten weeks. More than twenty years of experience guarantee authors the best possible service. To reach the goal of rapid publication at a low price the technique of photographic reproduction from a camera-ready manuscript was chosen. This process shifts the main responsibility for the technical quality considerably from the publisher to the authors. We therefore urge all authors and editors of proceedings to observe very carefully the essentials for the preparation of camera-ready manuscripts, which we will supply on request. This applies especially to the quality of figures and halftones submitted for publication. In addition, it might be useful to look at some of the volumes already published. As a special service, we offer free of charge LaTeX and TeX macro packages to format the text according to Springer-Verlag's quality requirements. We strongly recommend that you make use of this offer, since the result will be a book of considerably improved technical quality. To avoid mistakes and time-consuming correspondence during the production period the conference editors should request special instructions from the publisher well before the beginning of the conference. Manuscripts not meeting the technical standard of the series will have to be returned for improvement.

For further information please contact Springer-Verlag, Physics Editorial Department II, Tiergartenstrasse 17, D-69121 Heidelberg, Germany

Daniel Benest Claude Froeschlé (Eds.)

Impacts on Earth

Editors

Daniel Benest
Claude Froeschlé
OCA Observatoire de Nice
BP 4229
F-06304 Nice Cedex 4, France

Cataloging-in-Publication Data applied for.

Die Deutsche Bibliothek - CIP-Einheitsaufnahme

Impacts on earth / Daniel Benest ; Claude Froeschlé (ed.). - Berlin ; Heidelberg ; New York ; Barcelona ; Budapest ; Hong Kong ; London ; Milan ; Paris ; Santa Clara ; Singapore ; Tokyo : Springer, 1998
(Lecture notes in physics ; 505)
ISBN 3-540-64209-9

ISSN 0075-8450
ISBN 3-540-64209-9 Springer-Verlag Berlin Heidelberg New York

Typesetting: Camera-ready by the authors/editors
Cover design: *design & production* GmbH, Heidelberg
SPIN: 10644115 55/3144-543210 - Printed on acid-free paper

Editors' Preface

by Daniel Benest and Claude Froeschlé

Cosmic Impacts

At the beginning of the 19th century, astronomers discover the first asteroids ... and the academics recognize that stones may fall from the extra-atmospheric space. Half a century later, many other asteroids had been discovered, and Daniel Kirkwood found that there were gaps in their semi-major axis distribution, corresponding to mean motion resonances with Jupiter. As discovered in 1918 by Kiyotsugu Hirayama, some groups are apparent in the frequency distribution of the orbital elements of these minor planets, which he called *families*; Hirayama suggested that the origin of these families could come from a catastrophic break-up of a common parent body. Finally, since the 80's, modern celestial mechanics and dynamical planetology have begun to explain the processes that send asteroids on trajectories crossing the Earth's orbit. This will be developed in Part I of this book.

We know now that impact phenomena play an essential role in the formation of planets – and of other minor bodies orbiting the Sun – and keep a great influence during their evolution; impacts have therefore a non negligible importance in the history (particularly from the point of view of geology) of the Solar System. Part II describes some physical processes consecutive to the impact on a lithosphere, and Part III discusses the dating of ancient impacts on Earth together with some considerations about hazards due to space debris orbiting our planet.

Impacts cosmiques

Au début du XIXe siècle, les astronomes découvrent les premiers astéroïdes ... et les Académiciens reconnaissent que des cailloux peuvent tomber du ciel. Un demi-siècle plus tard, de nombreux autres astéroïdes avaient été découverts, et Daniel Kirkwood détecta des lacunes dans leur distribution spatiale en demi-grand axe, qui correspondent à des résonances en moyen mouvement avec Jupiter. En 1918, Kiyotsugu Hirayama mit en évidence certains groupements dans la distribution en fréquence des éléments orbitaux de ces petites planètes, qu'il nomma *familles*; Hirayama suggéra que l'origine de ces familles pourrait se situer lors du brisement catastrophique d'un corps "parent". Finalement, depuis les années 80, la mécanique céleste moderne et la planétologie dynamique commencent à comprendre les mécanismes qui envoient des astéroïdes sur des trajectoires qui croisent l'orbite de la Terre. Ce qui est développé dans la Partie I du présent ouvrage.

Nous savons maintenant que les phénomènes d'impact jouent un rôle essentiel dans la formation des planètes – et des autres petits corps qui orbitent autour du Soleil –, et conservent une grande influence au cours de leur évolution; les impacts ont donc une importance non négligeable dans l'histoire du Système Solaire (particulièrement du point de vue de la géologie). La Partie II éclaire certains processus physiques consécutifs aux impacts sur une lithosphere, et la Partie III expose les méthodes de datation des impacts terrestres anciens, ainsi que quelques considérations à propos des dangers que peuvent faire courir les débris spatiaux en orbite autour de notre planète.

The Goutelas School

The Spring School of Astronomy and Astrophysics of Goutelas has taken p[lace since 1977. Founded by Evry Schatzman, of the French Sciences Academy, the school is held annually at Goutelas Castle in the Forez country (Haute-Loire, France), under the patronage of the S.F.S.A. (Société Française des Spécialistes d'Astronomie) and with the financial support of the C.N.R.S. (the French "Centre National de la Recherche Scientifique") through its "Continuing Formation" department. The manager and the staff of the "Centre Culturel de Goutelas" contribute a lot to the success of the school, as they do their best to provide pleasant and friendly surroundings.

In the past, for the schools of 1989 and 1991 to and for other meetings, we had already proposed topics closely related to the dynamics of minor bodies in the Solar System, or about techniques used in dynamical planetology (see references). The 18th Goutelas School was held during 2nd-7th May 1994, and was devoted to the study of these impact phenomena, and particularly those affecting our planet, the Earth. This book is born from the courses given in this school, and the chapters have been updated.

Last but not the least, we were assisted by Monique Fulconis, with all her efficiency and kindness.

L'École de Goutelas

L'École de Printemps d'Astronomie et d'Astrophysique de Goutelas existe depuis 1977. Fondée par Evry Schatzman, de l'Académie des Sciences, l'École se tient annuellement au Château de Goutelas dans le Forez (Haute-Loire, France), sous le parrainage de la Société Française des Spécialistes d'Astronomie et avec le soutien financier de la Formation Permanente du Centre National de la Recherche Scientifique. Le succès de cette École tient pour une bonne part dans l'efficacité et la gentillesse de l'accueil du personnel du Centre Culturel de Goutelas.

Déjà, lors des Écoles de 1989 et 1991 – ainsi que lors d'autres réunions –, nous avions proposé des thèmes proches de la dynamique des petits corps du Système Solaire ou exposant des techniques utilisées en planétologie dynamique (cf refs.). La 18ᵉ École de Goutelas s'est tenue du 2 au 7 mai 1994, et fut dévolue à l'étude de ces phénomènes d'impact, et en particulier à ceux ayant affecté notre planète, la Terre. Ce livre en est issu, et les chapitres actualisés.

Enfin, nous souhaitons remercier Monique Fulconis, qui sut assurer le secrétariat et régler tous les détails matériels avec efficacité et gentillesse.

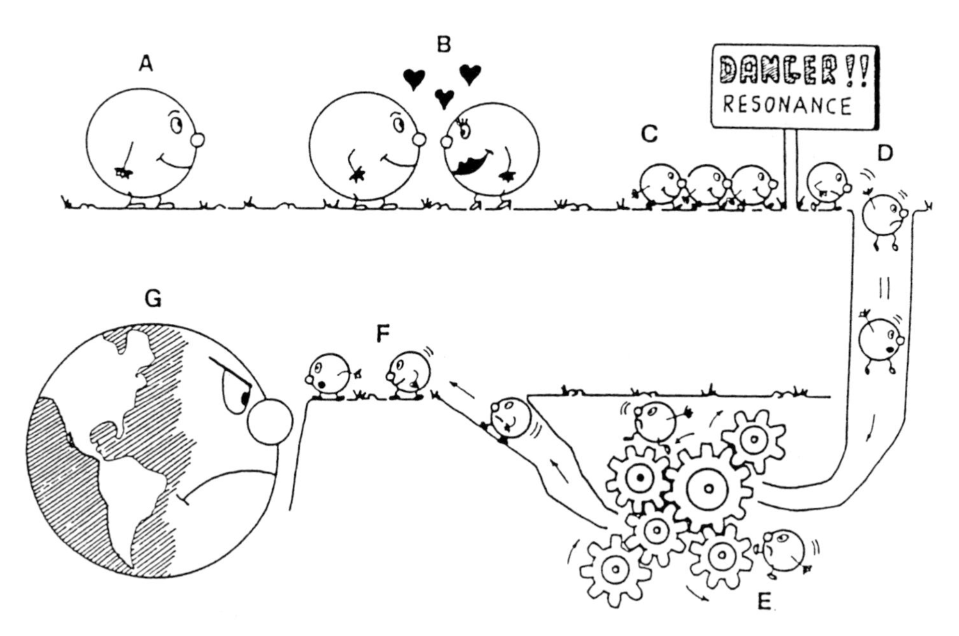

Fig. 1. Vincenzo Zappala, de l'Observatoire de Turin, illustre par ce dessin comment des fragments (C) issus d'une collision entre astéroïdes (A et B) peuvent être "piégés" dans une résonance (D) puis, par un mécanisme complexe (E), transférés dans les régions du Système Solaire interne (F) où ils risquent de rencontrer la Terre (G).
Vincenzo Zappala, from Turin Observatory, shows by this picture how fragments (C) produced during a catastrophic collision between asteroids (A and B) may be trapped in a resonance (D) then, by a complex mecanism (E), transfered in the inner regions of the Solar System (F), where there is a danger of collision with the Earth (G).

References

Benest, D., Froeschlé, C. (eds.) (1990): *Les méthodes modernes de la Mécanique Céleste [Modern Methods of Celestial Mechanics] (Goutelas 1989)*, Editions Frontières (C36).

Benest, D., Froeschlé, C. (eds.) (1992): *Interrelations between Physics and Dynamics for Minor Bodies in the Solar System (Goutelas 1991)*, Editions Frontières (C49).

Benest, D., Froeschlé, C. (eds.) (1994): *An Introduction to Methods of Complex Analysis and Geometry for Classical Mechanics and Non-Linear Waves (Chamonix 1993)*, Editions Frontières.

Benest, D., Froeschlé, C. (eds.) (1995): *Chaos and Diffusion in Hamiltonian Systems (Chamonix 1994)*, Editions Frontières.

Benest, D., Froeschlé, C. (eds.) (1998): *Analysis and Modelling of Discrete Dynamical Systems - with Applications to Dynamical Astronomy (Aussois 1996)*, Gordon and Breach, in press.

Contents

Part I Origin and Dynamics of Projectiles

Part III Terrestrial Impacts

List of Participants

Olivier BARALE
O.M.P. Observatoire de Toulouse, 14 avenue Edouard Belin,
F-31400 TOULOUSE (France)

Daniel BENEST
O.C.A. Observatoire de Nice, B.P. 4229, F-06304 NICE Cedex 4 (France)

Hervé CABOT
Laboratoire de Glaciologie du CNRS, 54 rue Molière, B.P. 96,
F-38402 SAINT-MARTIN d'HERES (France)

Alberto CELLINO
Osservatorio astronomico di Torino, Strada Osservatorio 20,
I-10025 PINO TORINESE (Italia)

Jacques CHAUVILLE
Observatoire de Meudon (DASGAL), 5 place Janssen,
F-92195 MEUDON Cedex (France)

Donald R. DAVIS
Planetary Science Institute, 620 North 6th Avenue, TUCSON
AZ 85705-8331 (USA)

Jean-Claude DOUKHAN
Lab. Struct. & Prop. Etat Solide, Univ. Sciences et Techniques de Lille
Bat. C6, F-59655 VILLENEUVE D'ASCQ Cedex (France)

Achim ENZIAN
Laboratoire de Glaciologie du CNRS, 54 rue Molière, B.P. 96,
F-38402 SAINT-MARTIN d'HERES (France)

Paolo FARINELLA
Universita di Pisa, Dipartimento di Matematica, Via Buonarotti 2,
I-56127 PISA (Italia)

Michel FESTOU
O.M.P. Observatoire de Toulouse, 14 Avenue Edouard Belin,
F-31400 TOULOUSE (France)

Christiane FROESCHLÉ
O.C.A. Observatoire de Nice, B.P. 4229, F-06304 NICE Cedex 4 (France)

Claude FROESCHLÉ
O.C.A. Observatoire de Nice, B.P. 4229, F-06304 NICE Cedex 4 (France)

Monique FULCONIS
O.C.A. Observatoire de Nice, B.P. 4229, F-06304 NICE Cedex 4 (France)

Jan-Martin HERTZSCH
O.C.A. Observatoire de Nice, B.P. 4229, F-06304 NICE Cedex 4 (France)

Patrick MARTIN
O.M.P. Observatoire de Toulouse, 14 avenue Edouard Belin,
F-31400 TOULOUSE (France)

Philippe MASSON
Laboratoire de Géologie dynamique interne, Univ. Orsay-Paris-Sud, bat. 509,
F-91405 ORSAY Cedex (France)

Patrick MICHEL
O.C.A. Observatoire de Nice, B.P. 4229, F-06304 NICE Cedex 4 (France)

André MIGAULT
Laboratoire de Combustion et de Détonique, E.N.S.M.A.,
B.P. 109, Site du Futuroscope, Chasseneuil du Poitou,
F-86960 FUTUROSCOPE Cedex (France)

Andrea MILANI
Universita di Pisa, Dipartimento di Matematica, Via Buonarotti 2,
I-56127 PISA (Italia)

Alesssandro MONTANARI
Osservatorio Geologico di Coldigioco, I-62020 FRONTALE DI APIRO (Italia)

Alessandro MORBIDELLI
O.C.A. Observatoire de Nice, B.P. 4229, F-06304 NICE Cedex 4 (France)

Edouard OBLAK
Observatoire de Besançon, 41 bis avenue de l'Observatoire, B.P. 1615,
F-25010 BESANÇON Cedex (France)

Daniel PETRINI
O.C.A. Observatoire de Nice, B.P. 4229, F-06304 NICE Cedex 4 (France)

Jean POHL
Institut für Allgemeine und Angewandte Geophysik, Theresienstrasse 41,
D-80333 MUENCHEN (Allemagne)

François POULET
Observatoire de Besançon, 41 bis avenue de l'Observatoire, B.P. 1615,
F-25010 BESANÇON Cedex (France)

Hans RICKMAN
Astronomiska observatoriet, P.O. box 515, S-75120 UPPSALA (Suède)

Robert ROCCHIA
Centre des faibles radioactivités, Avenue de la Terrasse,
F-91198 GIF-SUR-YVETTE Cedex (France)

Alessandro ROSSI
CNUCE, via Santa Maria 36, I-56126 PISA (Italia)

Urs SCHAERER
Lab. Géochronologie I.P.G., Univ. Paris 7, 2 Place Jussieu, tours 24-25, F-75251 PARIS Cedex 05 (France)

Jean SOUCHAY
128 rue Lecourbe, F-75015 PARIS (France)

Fabrice THOMAS
O.C.A. Observatoire de Nice, B.P. 4229, F-06304 NICE Cedex 4 (France)

Bruno VIATEAU
Observatoire de Bordeaux, B.P. 89, F-33270 FLOIRAC (France)

Vincenzo ZAPPALA
Osservatorio Astronomico di Torino, Strada Osservatorio 20, I-10025 PINO TORINESE (Italia)

I

Origin and Dynamics of Projectiles

Catastrophic Collisions in the Asteroid Belt – The Identification of Dynamical Families

Alberto Cellino[1] and Philippe Bendjoya[2]

[1] Osservatorio Astronomico di Torino, Strada Osservatorio 20, I-10025 Pino Torinese, Italia
[2] O.C.A. Observatoire de Nice, B.P. 4229, F-06304 Nice Cedex 4, France

Collisions catastrophiques dans la Ceinture des Astéroïdes : L'identification des familles dynamiques

Résumé. Les collisions jouent un role capital dans notre Système Solaire. On leur attribue la responsabilité d'évidences observationnelles comme la formation de la Lune, le basculement de l'axe des pôles d'Uranus, la formation des anneaux planétaires, l'intense cratérisation de la surface des planètes telluriques et l'ensemble des propriétés de la population des astéroïdes. La physique des collisions est une science complexe et ses modèles reposent sur des expériences de laboratoire au cours desquelles des cibles sont détruites par des projectiles supersoniques ou par des charges explosives. Les paramètres physiques comme la distribution des masses et des moments angulaires, ou les énergies mises en jeu, sont enregistrés. Hélas ces expériences limitent le champ d'action des modèles à cause de la taille des cibles (de l'ordre de quelques dizaines de centimètres) et des énergies mises en jeu.
Les astéroïdes offrent des expériences grandeur nature aux investigations des chercheurs. En effet, tant leur origine que leur histoire sont gouvernées par les collisions. Ainsi, les familles d'astéroïdes sont les vestiges de la destruction d'un astéroïde parent sous l'effet d'une collision catastrophique avec un autre astéroïde projectile. L'ensemble des fragments, qui sont eux-mêmes des astéroïdes orbitant autour du Soleil, sont les membres des familles. La connaissance précise et fiable du nombre de ces familles ainsi que de leur composition est donc capitale non seulement pour la modélisation des collisions catastrophiques sur des corps de quelques kilomètres à quelques centaines de kilomètres en taille, mais aussi pour la compréhension de l'origine et de l'évolution des astéroïdes.
Dans ce chapitre, nous présentons deux méthodes de détermination des familles basées sur des considérations dynamiques. Le résultat, fiable et tenant compte des fluctuations statistiques de ces classifications, servira de point de départ à une étude systématique des paramètres physiques des membres des familles. Il sera alors possible de reconstruire à la manière d'un puzzle tridimensionnel les planétésimaux à l'origine de la ceinture d'astéroïdes actuelle. Ces informations fourniront des contraintes très strictes, indispensables pour les modèles de formation du Système Solaire.

Abstract. Collisions play a capital role in the Solar System. Moon origin, tilt of Uranus pole axe, planet ring formation, heavy telluric planet surface craterisation, and the actual figure of the asteroid population are due to collisions. Unfortunately, models of collisions lay on laboratory experiments in which the involved targets and projectiles are centimeter sized objects. It is then difficult to apply scale laws to predict mass,

velocity or spin distributions of fragments resulting from a collision between astronomical bodies. Parameters such as the gravitational attraction between the different fragments are not taken into account in laboratory experiment due to the small size of the bodies.
The asteroid population is the best laboratory to study collisions between kilometer sized bodies. The origin and history of asteroids is closely linked to collision occurences, and asteroid families are the final product of a catastrophic break-up of a parent body. The members of an asteroid family are asteroids with a common origin and keeping evidences of the hyper-energetic collision they are issued from. The study of the members of different asteroid families is of great interest in order to stress the collisional models between kilometer sized bodies. Moreover the knowledge of the number of asteroid families will allow us to know the number of planetesimals in the early times of the Solar System formation and hence will permit to stress the models of Solar System formation. The first step of these studies is to determine the asteroid families and to have a list of family members on which later studies will be carried out. The aim of this chapter is to described two methods of asteroid family determination based on dynamical considerations. These totally independent methods allow to give a level of confidence to family members against chance fluctuations and so propose a reliable list of members for which physical and chemical (spectroscopy) parameters will be investigated.

1 Introduction

During the last decades the important role played by the occurrence of catastrophic collisions between the planetary bodies of our solar system has been progressively evidenced. Collisional events can explain a great deal of observational evidence, including the origin of our Moon, the anomalous tilt of the spin axis of Uranus, the formation of planetary rings, the intense craterization of planetary surfaces, and the overall properties of the asteroidal population.

The physics of the catastrophic break–up of solid bodies has long been widely investigated due to its importance for many practical (and also military) applications. However, this branch of physics is very complex, and it is very difficult to undertake a satisfactory fully–analytical approach, while any quantitative prediction based on numerical simulations requires a great amount of computing power (Melosh et al. 1992; Benz and Asphaugh 1994; Benz et al. 1994).

As a consequence, for a long time the main sources of information about catastrophic break–up phenomena have been laboratory experiments. Both techniques based on high velocity impacts (with velocities of the order of a few km/s) obtained by means of special guns, as well as techniques based on contact charges have been developed by different authors (see, e.g., Fujiwara et al. 1989, and references therein; see also Davis and Ryan 1990; Nakamura et al. 1992; Giblin et al. 1994). In turn, the experiments have triggered the development of semi–empirical models of catastrophic break–up phenomena (Paolicchi et al. 1989; Paolicchi et al. 1996).

Of course, the laboratory techniques have important limitations, mainly for what concerns the masses of the bodies involved. The typical sizes of the rocky bodies used in the experiments do not exceed 20 cm.

Such limitations are not present in the "experiments" carried out by Nature. This has been clearly recognized during the last decades by the people working in the field of planetary science. We presently know very well that in the Solar System there are many outstanding examples showing the relevance that catastrophic collisions have had in determining what we observe today.

Among these examples, the asteroid families are perhaps the subjects bearing the closest similarity with respect to the experiments performed in our laboratories. Families have been originated from impacts between couples of asteroids. The impact velocity of the projectile is mostly of the same order of the typical velocities reached in laboratory experiments, since the typical encounter velocities of objects orbiting in the asteroid belt range mostly between 2 and 6 km/s (Bottke et al. 1994). However, even if asteroids have fairly small masses with respect to the major planets and many of their satellites, they are enormously different with respect to the bodies used in our experiments. The mass of a 100 km -sized asteroid is about 10^{18} the mass of the biggest bodies used in laboratories. Of course, there are also other differences : Nature does not use any close chamber for her "experiments" with the asteroids. The fragments do not "fall in the ground", ready to be collected and analyzed, like in our laboratories. The fragments originated from the break-up of an asteroid are themselves asteroidal bodies, orbiting around the sun since the instant of their formation. The ejection velocities from the parent body are not so high with respect to its original orbital speed, and for this reason the fragments have orbits similar to that of the parent. On the other hand, even small orbital differences cause a rapid dispersion of the fragments in the asteroid belt. After a short time (some 10^5 years) the osculating orbits of the fragments and of course their actual positions at any given instant are very different. As a consequence, it is not easy to recognize the families in the asteroid belt.

On the other hand, the importance of the families is enormous both from the point of view of the study of the asteroids, as well as from the point of view of the physics of catastrophic break–up phenomena. For this reason, a big effort has been made recently in order to improve the techniques of asteroid family identification. The present paper is aimed at presenting an updated review of the state of the art in this field.

The real reason to develop refined techniques of family identification is not merely to produce a list of families for classification purposes. The real goal is to derive reliable information on the identity of the family members, in order to undertake a systematic *physical* investigation. What we need to know for each family, is a set of basic physical properties : the mass and spin rate distributions and the mineralogic compositions of the members, as well as some inherent properties of the original collisional event that produced the family, like the velocity distribution of the fragments, the mass of the parent body, and some plausible estimates of the kinetic energy of the impactor, the impact strength of the target, and the partition of the original impact energy. In turn, this information can trigger an enormous improvement of our present knowledge of both the physics of catastrophic break–up and the collisional history of the asteroidal population.

As for the first point, families are the outcomes of events that are far beyond the limits of energy that we can attain in the laboratory experiments. This makes them essential in order to test the present *scaling* theories. These are aimed at giving a good physical description of catastrophic break–up processes involving large bodies, starting from the available empirical evidence on the outcomes of laboratory experiments, and on the basis of the present theoretical knowledge. Breaking a km-sized object is very different with respect to breaking a 20-cm target, since the times of propagation of the compressive and tensile stresses caused by the impact within the impacted body are much larger; this makes some important differences on the overall behaviour of the break–up process. Moreover, in the case of asteroid-sized bodies, the effects of gravitation start to be very important. For these reasons, it is not easy to make refined theoretical predictions on the final result of the simultaneous interplay of many features that are different, or absent, in our laboratory experiments.

From the point of view of the asteroidal studies, the families are very important for many obvious reasons : generally speaking, their number puts some essential constraints about the overall process of collisional evolution of the asteroid belt. This is due to the fact that the amount of families presently recognizable depends upon the rate at which family-producing impacts occur in the belt, and upon the time that collisional erosion of the family members takes to make a family no longer identifiable.

On the other hand, the different members of each family are the different pieces of a three dimensional puzzle, and provide information on the internal structure of the parent body : possible differentiation, thermal history, original composition. In turn, this can provide some information about the physical conditions of the proto-planetary cloud.

Finally, families are also important from the point of view of the origin of meteorites. It has long been recognized that the main source of the meteorites hitting the Earth and the other terrestrial planets should be the asteroid belt. Recently, a big amount of work has been devoted to identify the main dynamical routes from the asteroid main belt to the inner solar system. Such routes are associated to the main *resonances*, both secular and of mean-motion, that are located in the belt (see Knežević and Milani 1994, for a review). The general idea is that impacts in the asteroid belt can inject fragments into some of the most important resonances. From there, these objects are quickly perturbed and achieve Mars and then Earth-crossing orbits. In this scenario, families are important, since they represent the most evident outcomes of energetic impacts, and at least some of them should have injected a fraction of the original members into the resonances. For what concerns the main mean-motion resonances with Jupiter associated to the Kirkwood gaps, a recent study by Morbidelli et al. (1995) shows that many of the most important families presently known have been presumably sources of Near Earth Objects in the past. In some cases, this can be particularly interesting, since the events associated to families are the only known events capable of injecting into the inner solar system some sizeable fragments, with diameters beyond 10 km. In this sense, they can explain the existence of NEAs like Eros and Ganymede, for which an alternative origin appears difficult to explain just due to their size.

A good example of the relevance of the family studies for the asteroidal science is provided by the results concerning the family associated to the large asteroid 4 Vesta. This family was first listed by Williams (1979, 1989) and then confirmed independently by Zappalà et al. (1990, 1994) and Bendjoya et al. (1991, 1993). On the basis of this evidence, Binzel and Xu (1993) undertook a spectroscopic campaign of the family. The aim was to check the reliability of this family by looking at the reflection spectrum of the proposed members. 4 Vesta was long known to be a unique example of differentiated asteroid, showing a basaltic surface clearly recognizable from the absorption bands around 1 and 2 μm. The results of the observations have fully confirmed the existence of the family, since the observed members have clearly shown the predicted basaltic composition. Due to its proximity to the important 3/1 mean-motion resonance with Jupiter, the event responsible of the formation of the Vesta family should also have injected a fraction of fragments in the zone of the terrestrial planets. This strengthens the hypothesis that Vesta is the parent body of the basaltic achondrites (eucrites).

This example shows that the field of family investigation can be particularly fruitful. In what follows we will focus our discussion on the most modern and efficient methods of family identification, since the possibility of having at disposal a set of families reliably identified is the necessary pre-requisite for any further physical development.

Historically, the identification of the families has long been a weak point, since until recently the big discrepancies between the differently proposed classifications made impossible any further investigation. For this reason, we will describe in some details the two most recent and independent methods that have been shown to be very "objective" and efficient, and, due to the excellent agreement of their results, have allowed to propose a list of very "robust" and reliable families, that are presently adopted by the most recent physical studies.

2 How to Identify Families

Reliably identifying the asteroid families is a difficult task. In this Section we shortly sketch the kind of observational evidence we have at disposal, and describe the general problems, dynamical and statistical, that have to be faced.

As quoted in the first Section, families are identifiable because the fragments of the collision are ejected at relatively low speeds compared to the orbital velocity of the parent body. This fact is important, since it puts some constraints on the kind of general event we are dealing with : the original collision must be energetic enough to completely shatter the target and disperse the fragments by overcoming its self-gravitation, but it cannot be so energetic to pulverize the target (because the fragments would not be observable) and/or to eject the fragments at very high velocities, since in this case the resulting orbits could be too dispersed since the beginning.

Even when the above energetic constraints are met, the family cannot be identified by looking at the present osculating orbits only. This is easily understood, taking into account that the orbital motion of the asteroids is strongly perturbed by the presence of the major planets, mainly Jupiter. As a consequence, the orbital elements that any asteroid has at a given instant (the osculating elements) are not constant in time, as they would be in the absence of gravitational perturbations. Instead, the osculating elements oscillate under the effects of the perturbations, and even a tiny difference of the oscillating frequencies leads with time to big differences of the orbital elements, even for objects that at a given instant may have very similar orbits, like the newly born members of a family.

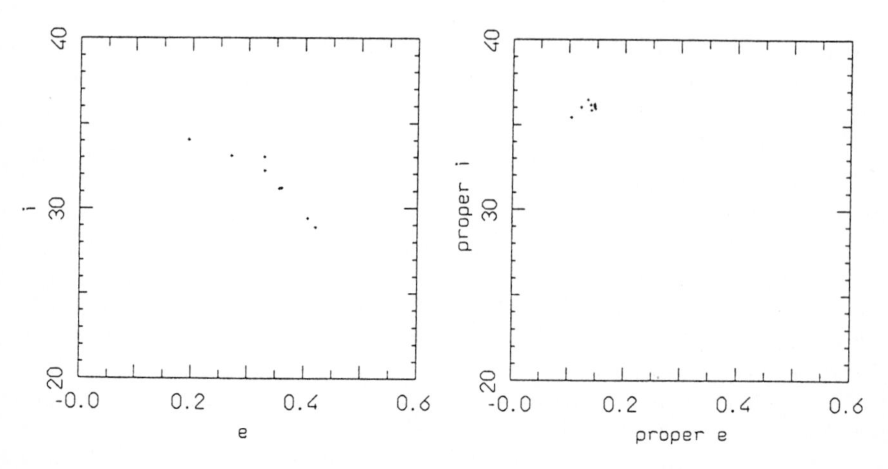

Fig. 1. Comparison between the osculating and the proper elements of some asteroids in an area around the asteroid Pallas.

As a consequence, due to the planetary perturbations, the osculating orbital elements of the members of a family do not keep the track of a common history, and we have to find other orbital parameters that are conserved along time. These parameters exist, and are the so-called *proper elements*. In the framework of a linear theory of the perturbations the proper elements are first integrals of the equations describing the temporal evolution of the orbital elements (see, e.g., Scholl et al. 1989). In the framework of the most refined and modern theories of the perturbations, where the perturbing function is expanded to higher degrees, proper elements can still be derived, although in this case they are quasi–integrals of the equations describing the variation of the orbital elements. On the other hand, the stability with time of the resulting proper elements has been extensively checked by means of numerical integrations, and has been found to be more than satisfactory over long time spans (for an extensive review of the most recent work on this subject, see Milani and Knežević 1994).

Due to their stability, the space of the proper elements a', e' and i' (proper semi-major axis, proper eccentricity and proper inclination, respectively) provides the natural environment in which asteroid families can be looked for. In this space, any anomalous condensation of points can be interpreted as the signature of a collisional event, since the proper elements of the members of any given family are generally very close to each other, and they remain constant over long time scales.

In order to illustrate why the proper elements provide the good information for asteroid family identification purposes, we show in Fig. 1 a comparison between the osculating and the proper elements of some asteroids in an area around the asteroid Pallas. This figure, provided by A. Morbidelli, shows how the indication of a possible common origin arises when we consider the proper elements distribution, whereas this possibility is much less evident in the osculating elements space.

Figure 2 shows the proper elements distribution of about 12,000 asteroids computed by Milani and Knežević according to an improved Yuasa theory of perturbation (see Milani and Knežević 1994, and references therein).

The inhomogeneity of the distribution is evident. One can observe empty areas which correspond to the Kirkwood gaps and populous regions with some very well contrasted concentrations. Due to the above considerations, such dense groupings can be viewed as the possible consequences of collisional events, i.e., they are the present family candidates. Of course, in a search for families it is necessary to develop reliable statistical methods, in order to assess the probability for any observed condensation to be due to pure chance. This is the most delicate part of the problem. In what follows, we will give a brief description of the most updated techniques that have been developed by different authors in the last years.

Intuitively, we tend to consider families as groups of orbits that are close enough to each other to make an overdensity with respect to the background. In order to avoid to be influenced by the subjective response of the pure eye, in recent years the general aproach has been made as "objective" as possible, through the development of techniques based on the use of statistical methods managed by the computer. In order to quantify the term " close enough" it is necessary first to define a metric in the proper elements space. The problem of the choice of a metric is not a trivial one. In the most recent approaches to this problem, the idea has been to exploit the well known relation between a sudden change of velocity of a body orbiting around the Sun, and the corresponding variation of its orbital parameters (Gauss equations). Indeed, such an approach seems particularly well suited to the scenario of family formation. As a consequence, as we are going to show, the adopted distance in the proper elements space has the dimensions of a velocity, according to the most recent family searches available in the literature (Zappalà et al. 1990; Bendjoya et al. 1991; Bendjoya 1993; Zappalà et al. 1994, 1995; for a general review on this topic, see also Milani et al. 1992).

Assuming that a sudden change of velocity δv is imparted to a body in heliocentric orbit, and that δv is small with respect to the orbital motion, the Gauss equations can be expressed in the form :

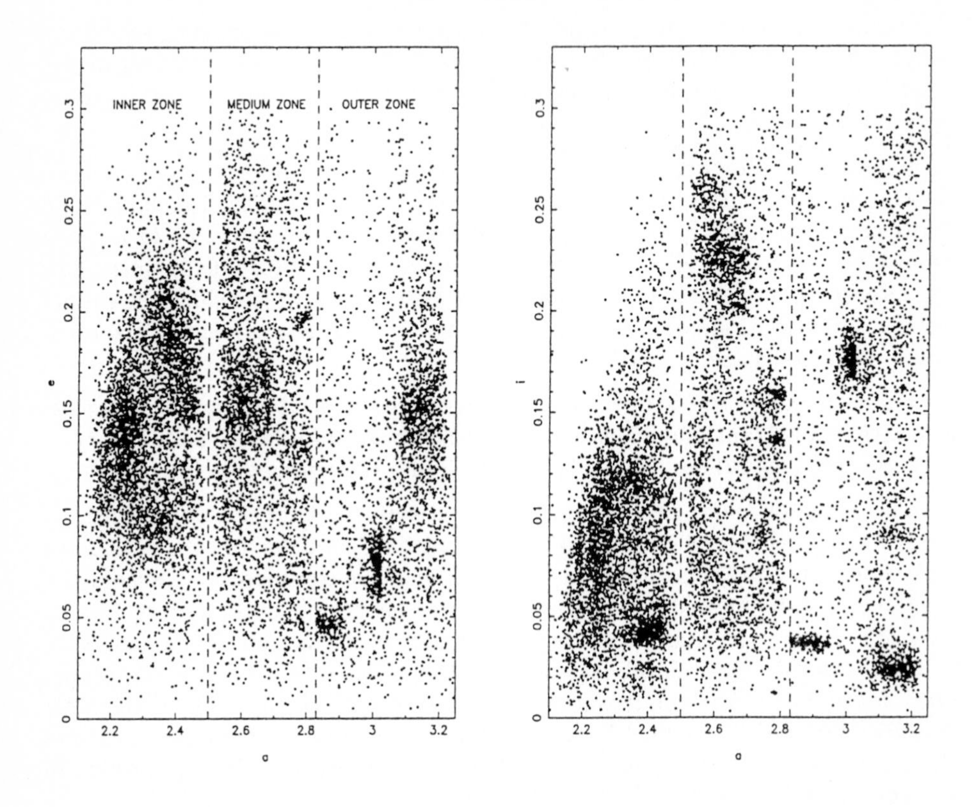

Fig. 2. The proper elements distribution of about 12,000 asteroids computed by Milani and Knežević according to an improved Yuasa theory of perturbation (see Milani and Knežević 1994, and references therein).

$$\delta a = \frac{2}{n(1-e^2)^{1/2}} \left[(1 + e\cos\nu)\delta v_T + e\sin\nu.\delta v_S \right] \tag{1}$$

$$\delta e = \frac{(1-e^2)^{1/2}}{n\,a} \left[\frac{e + 2\cos\nu + e\cos^2\nu}{1 + e\cos\nu} \delta v_T + \sin\nu.\delta v_S \right] \tag{2}$$

$$\delta i = \frac{(1-e^2)^{1/2}}{n\,a} \frac{\cos(\nu+\omega)}{1+e\cos\nu} \delta v_W \tag{3}$$

where δa, δe and δi are the resulting changes in the semi-major axis, eccentricity and inclination, respectively; δv_S, δv_W and δv_T are the components of δv along the radial component of the ejection velocity, the normal to the orbit plane component, and the component on $S \times W$ direction, respectively; n the mean motion, and ν and ω are the argument of the perihelion and the true anomaly, respectively, at the instant of the velocity change (the impact). By neglecting the terms proportional to the eccentricity, these equations can be simplified as :

$$\delta a = \frac{2\delta v_T}{na} \tag{4}$$

$$\delta e = \delta v_S \ \frac{\sin(\nu)}{na} + 2\delta v_T \ \frac{\cos(\nu)}{na} \tag{5}$$

$$\delta i = \delta v_W \ \cos(\omega + \nu) \tag{6}$$

The idea is now to express the relative velocity between two objects as a function of their differences in the orbital elements. In order to give an order-of-magnitude estimate of the velocity increment causing the separation of the two orbits one can define the relative velocity as follows :

$$\delta v = na\sqrt{k_1 \left(\frac{\delta a'}{a'}\right)^2 + k_2 \delta e'^2 + k_3 \delta i'^2} \tag{7}$$

where k_1, k_2, k_3 are of order of unity. Since ω and ν are unknown angles, a simple idea is to square Eqs. 4, 5, 6 and to average them over these angles. Substituting in Eq. 7 the δ(proper elements) by the right members of Eqs. 4, 5, 6 yields to :

$$\delta v = \sqrt{A < \delta v_T^2 > + B < \delta v_S^2 > + C < \delta v_W^2 >} \tag{8}$$

with $A = (4k_1 + 2k_2)$, $B = k_2/2$, $C = k_3/2$.

Unfortunately, it is impossible to obtain $A = B = C = 1$, which would correspond to give an equal weight to the three velocity components, with $k_1, k_2, k_3 > 0$. For this reason, Zappalà et al. (1990), have chosen the following set of coefficients, which define their *standard* metric : $k_1 = 5/4$, $k_2 = 2$, $k_3 = 2$. This coresponds to giving a higher weight to the δv_T component (A=9, B=1, C=1). This is consistent with the fact that the proper semi-major axis is the most accurate and stable proper element (see Milani and Knežević 1993). This standard metric has been adopted by Zappalà et al. (1990) and also in all the subsequent papers published by the Torino and Nice groups, quoted above. In all these papers, moreover, in order to test the sensitivity of the results upon the choice of the metric, another alternative metric has been adopted, defined according to the following choice of the coefficients : $k_1 = 1/2$, $k_2 = 3/4$, $k_3 = 4$. Such a choice gives a higher weight to the v_W component (related to the differences in the inclination).

As a final remark in this Section, we can note that the Gauss equations are in principle valid in the *osculating* elements space. However, Brouwer (1951) has shown that they can are still valid in the *proper* elements space in the framework of the linear theory of perturbations. This is still true also when proper elements computed on the basis of most refined theories of perturbation are chosen, as has been checked by means of extensive numerical integrations (Bendjoya et al. 1993) showing that the three-dimensional structure of a family in the osculating elelments space is closely reproduced when the computation of the proper elements for the same objects is carried out according to the theory of Milani and Knežević.

In the present Section we have shown that any "objective" technique of asteroid family identification is based on an analysis of the mutual distances of the asteroids in the proper elements space. The main problem, at this point, is to introduce reliable statistical methods, in order to analyze the proper elements distribution of asteroids, and to discriminate between the condensations that could be due to chance, and the groupings for which the only reliable explanation is based on a collisional origin. In the following Section we briefly describe the two most recent methods that have been used in the last years.

3 Two Independent Clustering Methods

It can be useful to begin this Section with a brief historical review. The term "family" was first used by Hirayama in 1918, when he observed in the orbital element space of the asteroids known at that epoch some condensations of orbits which were too big to be due to chance. Hirayama wrote : *"On examining the distribution of the asteroids with respect to their orbital elements, we notice condensations here and there. In general, they seem to be due to chance. But there are some which are too conspicuous to be accounted for by the laws of probability alone"*. The three major Hirayama families were Themis (22 members), Eos (21 members) and Koronis (13 members).

Since this pioneering work many authors have proposed family lists based on different sets of data. Table I shows a list of the works from 1918 to 1989 with the names of the authors, the number of asteroids considered in each search, nd the number of resulting families, and the adopted method of identification with the corresponding statistical test of reliability, if any.

The main conclusion of this comparison concerns the large range in the number of families proposed by the different authors. Such a large discrepancies have long been responsible of a general skepticism about the reliability of the different family lists. In some cases (Chapman 1989) also an analysis of the taxonomic types of the proposed family members showed that they could not be reconciled with the current ideas about the cosmochemistry of asteroids.

Of course, the discrepancies between the family lists reported in Table 1 could be due to the differences between the data-sets used by different authors. However, it seems difficult to explain how weak fluctuations in the number of analyzed asteroids could produce such huge discrepancies in the number of resulting families, at least in some cases (see, e.g., the 72 and 117 families found using data of 2125 and 2065 objects, respectively). As for the fact that different authors have used data-sets of asteroid proper elements computed by means of different techniques, Zappalà et al. (1992) have shown that such differences are not the main source of disagreement, by applying the same method of statistical identification to some of the data-bases of proper elements used by different authors.

Therefore, the only good explanation of the quoted discrepancies concerns the differently adopted methods of identification. As amatter of fact, many of the methods adopted in the past were based on a visual determination, which is by definition subjective and not reproducible.

For this reason, two automated statistical methods of family identification have been developed since 1990, and have caused a drastic improvement of the

Table 1.

Author(s)	Year	Numbered Ast. used	Number of Families	Method of detection	Stat Check
Hirayama	1933	1223	5	visual	no
Brouwer	1951	1563	28	visual	no
Arnold	1969	1735	37	boxes	yes
Linblad and Southworth	1971	1735	34	Southworth criterion	not applic.
Carusi and Massaro	1978	1861	15	z^2	yes
Williams	1979	1796	104	visual	yes
Kozai	1979	2125	72	visual	no
Williams	1989	2065	117	visual	yes

situation, giving results in a very good agreement. Moreover, being based on procedures fully managed by the computer, these two methods are fairly fast, fully reproducible, and "objective". Moreover, the performances of these methods can be directly tested by means of numerical simulations, as we will see later.

These two methods have been widely discussed in a series of papers : (Zappala et al. 1990, 1994, 1995; Bendjoya et al. 1991; Zappalà and Cellino 1992; Bendjoya and Cellino 1992; Bendjoya 1993). In the next subsections we only recall the main philosophy of both approaches.

3.1 The Hierarchical Clustering Method

The single-linkage hierarchical clustering method is a well known technique of multivariate data analysis (see Murtagh and Heck 1987), that consists in agglomerating subsequently the closest clusters. Being given a definition of distance in the proper elements space, all the mutual distances between the objects are computed, then at the first step the two closest objects i and j are identified and agglomerated, i.e., they are replaced by a unique object, for which the distance between it and any other object k is defined as the *minimum* between the distance (i, k) and the distance (j, k). All the distances are then updated and the same step is repeated as long as two objects survive. It is then possible to build a dendrogram giving for any value of distance the number of clusters existing, and the identity of their members. This allows to represent the results by means of suitable plots, the so called *stalactite diagrams*, in which the number of groupings and their numerical consistence are plotted for different values of distance. Families are then defined by comparing this diagrams with anal-

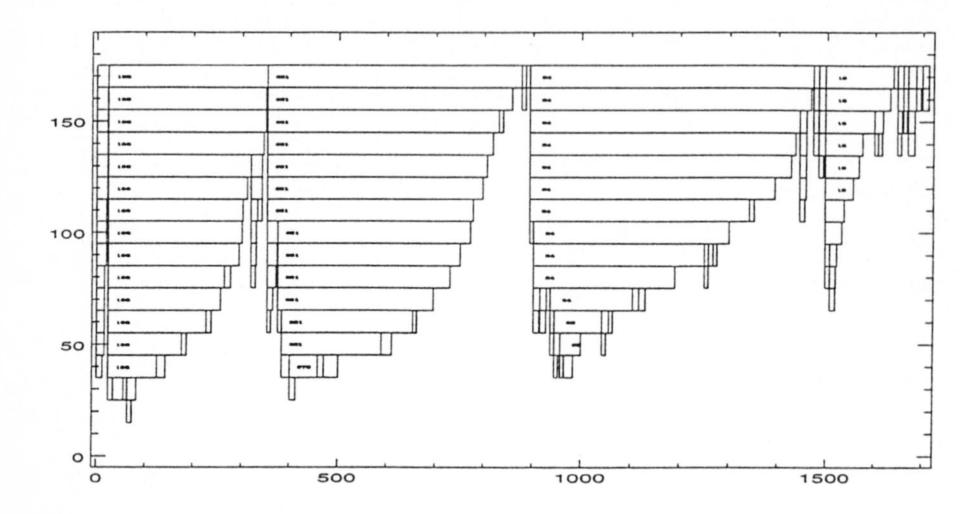

Fig. 3. Typical stalactite diagram referring to a zone of the main belt, with the critical distance value at 150 m/s.

ogous ones referring to a quasi random distribution of elements matching the large scale distribution of the real objects. The idea is to find a critical value of distance beyond which the probability that random agglomerations can have a given numerical consistence is fully negligible (see Zappala et al. 1990).

Figure 3 shows, as an example, a typical stalactite diagram referring to a zone of the main belt, with the critical distance value at 150 m/s.

3.2 The Wavelet Analysis Method

The wavelet transform cluster analysis is a density method based on the computation of a wavelet coefficient on each node of a network superimposed to the set of data. The wavelet coefficient can be seen as an indicator of proximity. The wavelet transform acts as a zoom and evidences structures which have the same size as the studied scale. By applying the wavelet transform to a quasi random distribution, built in the same way as in the hierarchical clustering technique, it is possible to define a threshold that quantify the risk that a detected structure is due to chance. The analysis is performed for a set of scales and structures appear to fit into each other from small to large scales. A criterion based on the philosophy of the wavelet transform is then applied in order to cut the hierarchy and to define families.

Both the techniques presented in this and the above Subsection are conceptually simple, allow to analyze quickly the available data-sets, and can be easily improved by changing some parts of the procedures if better criteria are suggested by some kind of evidence. As an example, different metric definitions can be used and quickly applied to the same data-set, different levels of detection can be envisaged and different sets of data can be fastly compared. Comparisons between the results of analyses in which different metrics are used, or the nominal proper elements are changed according to their nominal uncertainty, an lead to define some robustness parameters which estimate the real reliability of any detected family. Moreover, the sensitivity of the family membership to the level at which the hierarchy is cut is another information on its stability against the local background. In the framework of the Hierarchical Clustering method, as well as in that of the Wavelet Analysis method different nomenclatures have been introduced in order to indicate groupings of stronger or weaker reliability. In particular, both methods have allowed to evidence the existence of two main kinds of typologies among the most reliable families. The so-called *clans* are large groupings for which the number of members is very sensitive to the hierarchical cutting level; they can be the products of very energetic collisional events, or even of multi-generation collisions, in which some of the members of an older family can have been subsequently shattered by collisions, leading to an overlapping of the fragments in the proper elements space (see also Cellino et al. 1991). *Cluster* are instead very sharp and dense groupings, for which the membership is fairly insensitive upon the cutting level. Maybe, they could be the outcomes of most recent collisions, although any hypothesis must be checked by an extensive physical analysis. In particular, Farinella et al. (1992) have suggested to introduce a new nomenclature, in which the word "family" should be used only in the cases in which there is a concurring evidence of a collisional origin, coming from both a statistical and a physical investigation (mineralogic composition of the members, size and velocity distributions of the members, etc.). From this point of view, the best case of "family" known today is the one of Vesta. Due to the rapid increasing of available physical data on family members (Binzel et al. 1994), the situation should improve very much in the near future.

4 A "Quality Check"

Before giving some results obtained by the two methods described above, it can be interesting to show how the performances of these two procedures can be easily tested by means of numerical simulations. In so doing we take fully profit of the advantages offered by automated techniques, in which most of the work is caried out by the computer.

The general idea is to apply the two techniques to situations in which there is an *a priori* knowledge of the existence of a given family formed by a given number of members, and to compare the results of the two methods with the real situation. This can be done easily, if the "real" family is simulated one, generated numerically on the basis of suitable physical criteria. Such an exercise has been performed by Bendjoya et al. (1993). Here, we briefly describe the procedures applied in the above paper.

The following procedure has been applied to generate the artificial families.

Let V_{e_j}, V_∞, V_{esc} and V_{min} be, respectively, the ejection velocity at the instant of break-up, the corresponding velocity at infinity, the escape velocity from the parent body and the minimum ejection velocity, the latter being a parameter which characterizes the violence of the break-up. The following relations hold :

$$V_\infty = \left(V_{e_j}^2 - V_{esc}^2\right)^{1/2} \Rightarrow V_{e_j} = (V_\infty^2 + V_{esc}^2)^{1/2}$$

$$dV_{e_j} = V_\infty(V_\infty^2 + V_{esc}^2)^{-1/2} dV_\infty$$

For a homogeneous and spherical asteroid, having density ρ and radius R, the escape velocity is given by

$$V_{esc} = \sqrt{\frac{8}{3}\pi\rho G R}$$

(G being the universal constant of gravitation). Laboratory experiments on catastrophic breakups, indicate that the V_{e_j} distribution of the fragments can be fitted by a power law :

$$dN(V_{e_j}) = C V_{e_j}^{-\alpha} dV_{e_j}$$

where $dN(V_{e_j})$ gives the number of fragments having an ejection velocity between V_{e_j} and $V_{e_j} + dV_{e_j}$.

If the minimum ejection velocity of the fragments, V_{min}, does not exceed V_{esc}, it is evident that not all the fragments can reach infinity, and the minimum of the distribution of V_∞ is 0, corresponding to the fragments ejected with $V_{e_j} = V_{esc}$. On the other hand, if $V_{min} > V_{esc}$ there are no fragments with a V_∞ less than $(V_{min}^2 - V_{esc}^2)^{\frac{1}{2}} = V_{cutoff}$. Thus, the distribution of V_∞ is given by :

$$dN(V_\infty) = C V_\infty (V_\infty^2 + V_{esc}^2)^{-\frac{\alpha+1}{2}} dV_\infty$$

if $(V_{min} < V_{esc})$, or :

$$\begin{cases} dN(V_\infty) = & C V_\infty (V_\infty^2 + V_{esc}^2)^{-\frac{\alpha+1}{2}} dV_\infty \quad \text{if} \quad V_\infty > (V_{min}^2 - V_{esc}^2)^{\frac{1}{2}} \\ \qquad\quad = & 0 \qquad\qquad\qquad\qquad\qquad\quad\; \text{if} \quad V_\infty < (V_{min}^2 - V_{esc}^2)^{\frac{1}{2}} \end{cases}$$

if $(V_{min} > V_{esc})$.

The constant C is determined by normalization :

$$1 = C \int_{V_{lim}}^{\infty} V_\infty (V_\infty^2 + V_{esc}^2)^{-\frac{\alpha+1}{2}} dV_\infty$$

with $V_{lim} = 0$ if $V_{min} < V_{esc}$ and $V_{lim} = V_{cutoff}$ if $V_{min} > V_{esc}$. Let $n(v_\infty)$ be the fraction of fragments having a velocity at infinity less than v_∞ :

$$n(v_\infty) = \int_{V_{lim}}^{v_\infty} dN(V_\infty)$$

since $n(V_{lim}) = 0$

$$n(v_\infty) = (V_{lim}^2 + V_{esc}^2)^{\frac{\alpha-1}{2}} \left[1 - (v_\infty^2 + V_{esc}^2)^{\frac{1-\alpha}{2}}\right]$$

and finally :

$$v_\infty = V_{lim}\sqrt{(1 - n(v_\infty))^{\frac{2}{1-\alpha}} - 1}$$

Once that the osculating elements of a parent body M_0, ω_0, Ω_0, a_0, e_0, i_0 (mean anomaly, argument of perihelion, longitude of node, semi major axis, eccentricity and inclination, respectively) are fixed, its coordinates and velocity at the instant of break-up $(x_0, y_0, z_0, v_{x_0}, v_{y_0}, v_{z_0})$ in the cartesian phase space can be computed. N_f fragments are generated by taking randomly N_f values of $n(V_\infty)$ in the interval $[0, 1]$, so the N_f values of the module of ejection velocity at infinity are obtained for the fragments. Figure 4 shows the number of fragments with a velocity at infinity between V_∞ and $V_\infty + dV_\infty$ for two simulated families of 1000 fragments, with the fitting power law obtained from laboratory experiments : Fig. 4a for $V_{min} < V_{esc}$, Fig. 4b for $V_{min} > V_{esc}$.

The directions of the velocities at infinity obtained in this way are also randomized, leading to the velocity components $(v_{x_j}, v_{y_j}, v_{z_j})$ for $j = 1$ to N_f . Adding these velocity components to those of the initial parent body (v_{x0}, v_{y0}, v_{z0}), it is then possible to derive the osculating elements of each fragment $(M_j, \omega_j, \Omega_j, a_j, e_j, i_j)$ at the moment of break-up.

According to the results of laboratory experiments, α has been assumed equal to 3.25; R has been chosen to be 50 km, a realistic size for a parent body susceptible to create a substantial number of sizable fragments. Fixing $\rho = 2.5\ 10^3$ kg/m^3, which corresponds to silicate materials which are known to be a common constituent of a large number of asteroids, V_{esc} turns out to be 60 m/s.

In order to test the ability of both methods to detect structures more or less spread in backgrounds of varying density, different simulations, were performed, assigning to V_{min} the values 50 m/s, 100 m/s, 150 m/s and 200 m/s. In each case, the family so created was plunged in different backgrounds of 300, 600 and 900 elements. These backgrounds have been randomly created in a zone, far from the most important mean motion resonances, defined by $2.6 < a < 2.8$ A.U., $0 < e < 0.2$, $0 < \sin i < 0.2$, the three other osculating elements being taken randomly in the interval $[0, 2\pi]$. In each case, a single family composed by 51 members was created (a "parent body" belonging to the background plus 50 fragments). Figure 5 shows the projections of the "extreme"cases (Fig. 5a for $V_{min} = 50$ m/s and Fig. 5b for $V_{min} = 200$ m/s) in both the a-e and a-$\sin i$ planes.

Finally tests on the ability of the methods to separate close families have been performed. Two families : A (50 members) and B (30 members); have been generated with a non empty intersection in the first case and with tangential boundaries in the second case.

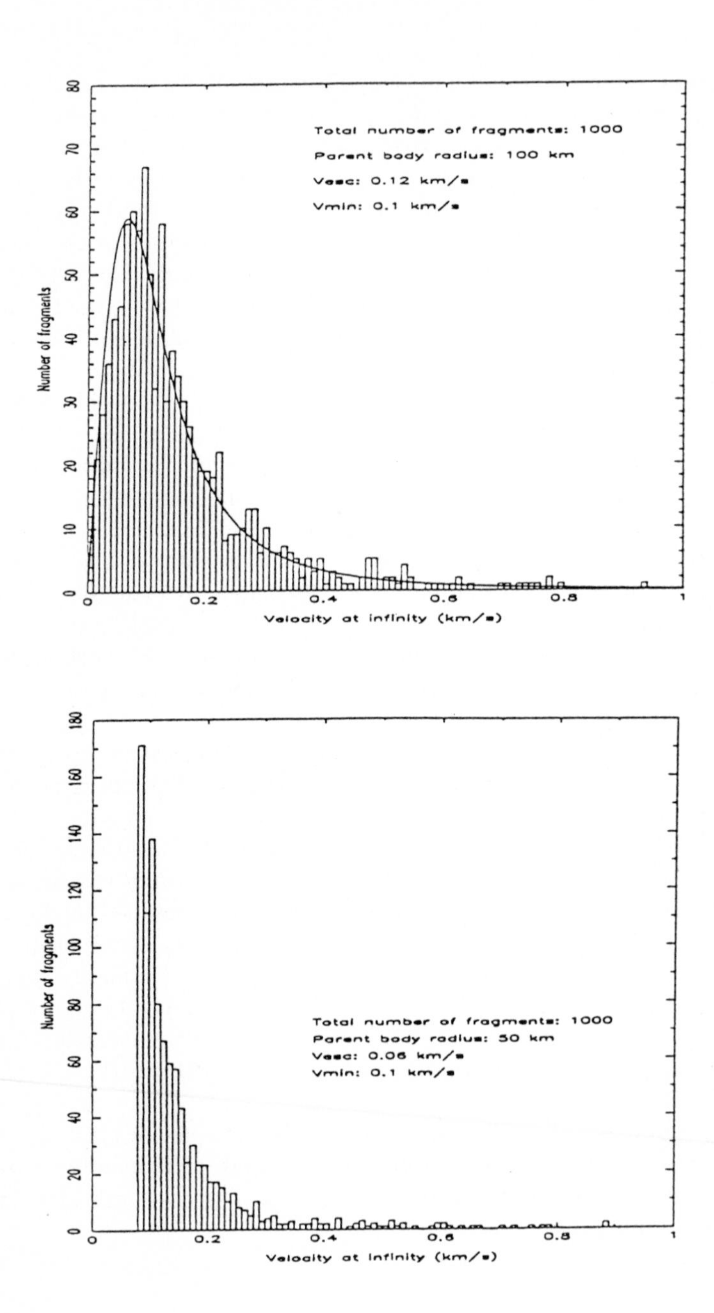

Fig. 4. Number of fragments with a velocity at infinity between V_∞ and $V_\infty + dV_\infty$ for two simulated families of 1000 fragments, with the fitting power law obtained from laboratory experiments: Fig. 4a for $V_{min} < V_{esc}$, Fig. 4b for $V_{min} > V_{esc}$.

Table 2. BG : number of asteroids in the background; **N** : number of asteroids rediscovered in the family; I : number of interlopers.

BG	V_{min} (m/s)							
	50		100		150		200	
	N	I	N	I	N	I	N	I
				HCM				
300	**50**	*1*	**46**	*1*	**44**	*5*	**10** **4** **8**	*2* *2* *0*
600	**50**	*7*	**46**	*7*	**46**	*14*	**13** **13**	*5* *8*
2 fam inter			**46A** **25B**	*10*				
2 fam tang			**45A** **24B**	*5* *1*				
900	**50**	*2*	**47**	*5*	**40**	*5*	**9** **5**	*2* *3*
				WAM				
300	**50**	*1*	**47**	*1*	**41**	*3*	**37**	*9*
600	**49**	*3*	**46**	*4*	**40**	*9*	**37**	*24*
2 fam inter			**69** **(45A,24B)**	*7*				
2 fam tang			**68** **(43A,25B)**	*5*				
900	**49**	*4*	**46**	*7*	**36**	*11*	**29**	*27*

The results are summarized in Table 2.

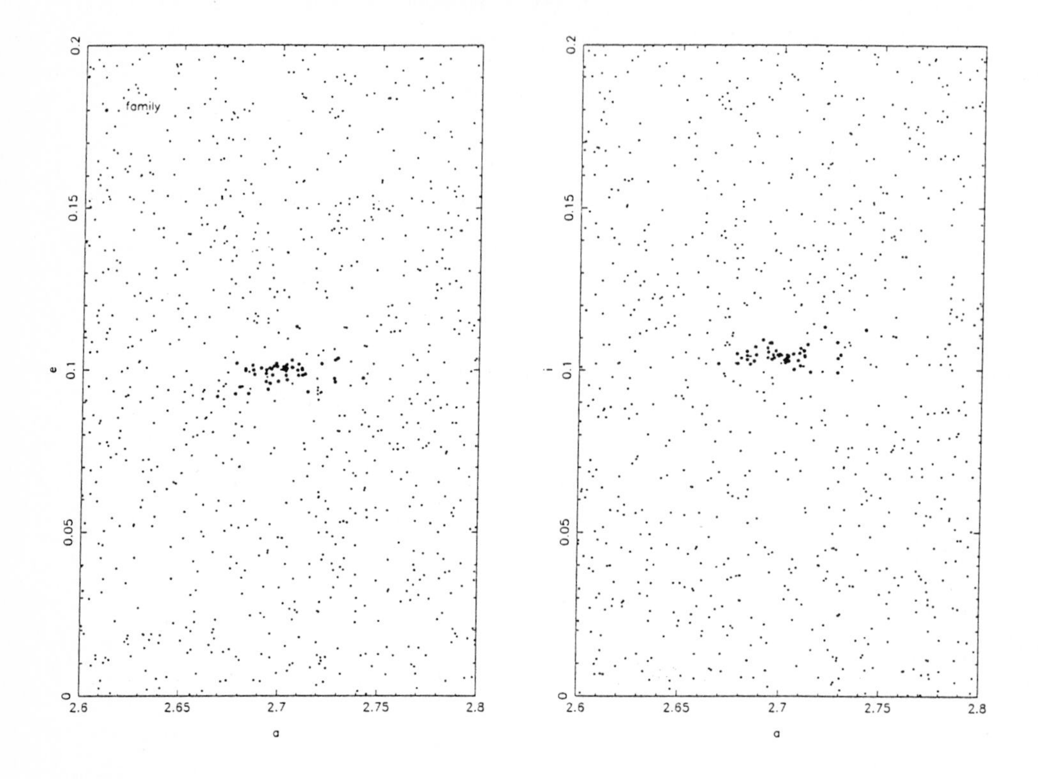

Fig. 5a. Projections of the "extreme" case $V_{min} = 50$ m/s in both the a-e and a-$\sin i$ planes.

The results of the different tests that have been performed have led us to consider these two independent methods as reliable. Since 1990, both the HCM and the WAM have been applied to increasingly larger data-sets of asteroid proper elements, and have given results in a very good mutual agreement, as can be seen in the quoted original papers, as well as in Zappalà and Cellino (1994). In the next section, we will focus in the most recent, and still unpublished family search carried out by Zappalà et al. (1995), who have applied separately both methods to the biggest data-set of proper elements ever used, including more than 12,000 objects, provided by Milani and Knežević.

5 The Most Recent Results

The last asteroid sample used includes 12,487 bodies, having proper semimajor axes (a), eccentricities (e) and sine of proper inclinations ($\sin I$) in the ranges 2.065 AU $< a <$ 3.278 AU, $0 < e < 0.3$ and $0 < \sin I < 0.3$. Thus we have searched the whole main asteroid belt (but not Trojans, Hildas, Cybeles and Aten–Apollo–Amors) with the exception of the zones at high inclinations and/or

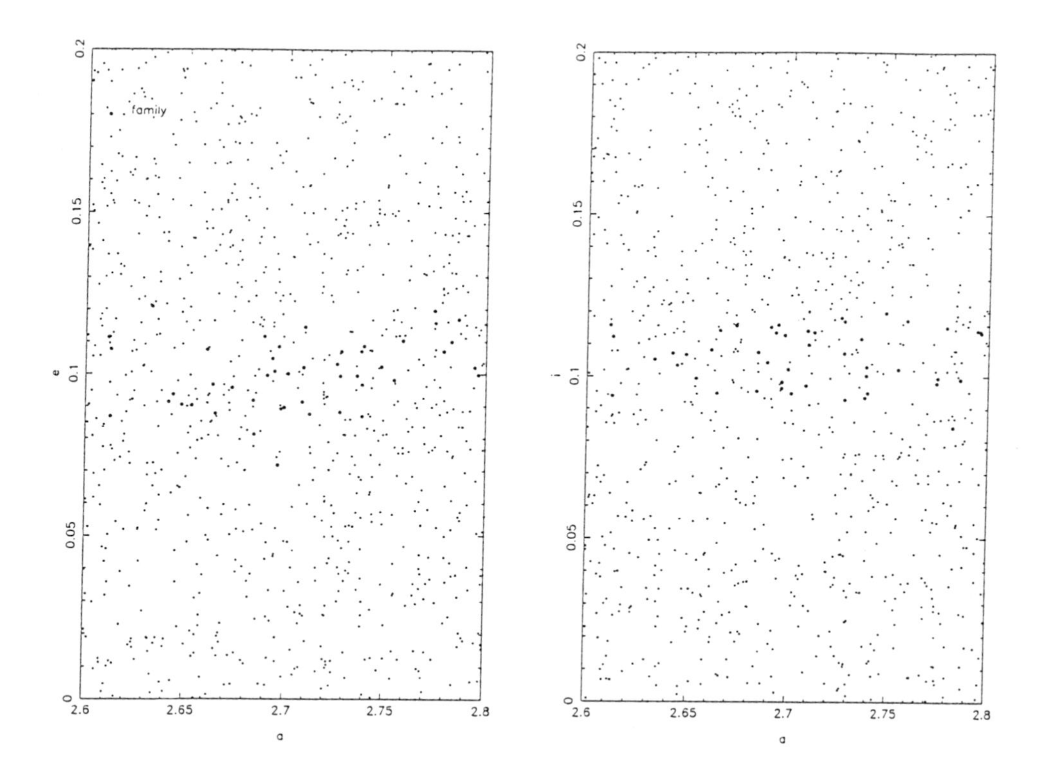

Fig. 5b. Projections of the "extreme" case $V_{min} = 200$ m/s in both the a-e and a-$\sin i$ planes.

eccentricities, for which different proper elements need to be used (Lemaître and Morbidelli 1994; we plan to work on these objects soon). As in Bendjoya (1993) and Zappalà et al. (1994), we have divided the main asteroid belt in three semimajor axis zones, in order to accomplish the family search in a more efficient way : an inner zone between the 4 : 1 and the 3 : 1 mean motion resonances with Jupiter; an intermediate zone between the 3 : 1 and the 5 : 2 Kirkwood gaps; and an outer zone, between the 5 : 2 and the 2 : 1 gaps .

Our sample includes 4616 asteroids numbered up to the beginning of 1993 and 7871 unnumbered asteroids. Whereas both Bendjoya (1993) and Zappalà et al. (1994) included in their sample only about 2200 unnumbered asteroids observed at more than one opposition, we are now including also more than 5000 bodies whose orbits have been derived on the basis of observations performed at one opposition only. These orbits were selected using the covariance matrices of the orbital elements computed by Muinonen and Bowell (1993) to assess their precision.

The whole set of the results of this analysis heve been published by Zappala et al. (1995). We propose here a partial result comparison in the next tables. The groupings having a significant number (N_{com}) of common members are identified with each other, although often they do not share the same name in the two lists. When such identifications are possible, the table also shows the percentages of objects found by each of the two techniques in the overall pool of members. Of course, the best agreement corresponds to high values for both percentages; when one of them is low, this means that a technique finds only a small subcluster of a larger grouping identified by the other one. The last columns of the tables and the footnotes provide some more information on the results of the comparison work.

The typographic code is the following: **"bold face"** for the most robust groupings i. e. those which are the least sensitive to the metric, to the uncertainty in the proper element and to the level of detection; "roman" for the groupings which fails one of the upper mentioned tests and hence whose membership is level detection dependent; *"italic"* for the least robust and hence more spurious detected groupings.

The comparison shows that almost all the most significant, robust and populous families, and also several smaller ones, are found by the two methods with well–overlapping memberships.

The most striking discrepancies concern essentially the smallest or/and the least robust detected groupings. A larger number of small families (with less than 20 members) are identified by WAM, especially in the intermediate zone, but are classified by HCM as clumps or not found at all; in several cases neighbouring groupings are merged by one technique, but not by the other.

Overall, our assessment of this comparison between the results of the two (completely independent) clustering techniques is that, on one hand, in a number of physically interesting cases the family memberships can now be considered as established in a very reliable way; but that, on the other hand, no "completely objective" family list can be derived by statistical methods, owing to the different limitations and ambiguities intrinsic in the data, which every searching technique can sort out only in a partial and specific way. Clearly, dynamical data must be supplemented by evidence from different sources (e.g., sizes, albedos, spectrophotometry) to derive unequivocal conclusion in the problematic cases.

Table 3. Analysis of the correspondance between the significant groupings found in the inner zone by the hierarchical clustering method (HCM) and by the wavelet analysis method (WAM). For each corresponding grouping, the Number N_{com} of common objects, as well as the percentages with respect to the total number of objects found by both methods, are listed.

Id(HCM)	N_{HCM}	Id(WAM)	N_{WAM}	N_{com}	$\%_{HCM}$	$\%_{WAM}$
8 Flora	604	**43 Ariadne**	575	434	71.9	75.5
44 Nysa	381	**135 Hertha**[3]	374	300	78.7	80.2
4 Vesta	231	**4 Vesta**	242	187	81.0	77.3
20 Massalia	49	20 Massalia	45	33	67.3	73.3
163 Erigone	45	**163 Erigone**	49	42	93.3	85.7
254 Augusta	19	(*)	–	–	–	–
752 Sulamitis[1]	7	752 Sulamitis	18	7	100.0	38.9
1454 Jerome[1]	7	**79 Eurynome**	107	7	100.0	[4] 6.5
4788 Simpson[1]	7	**43 Ariadne**	575	4	57.1	0.7
3038 Bernes[1]	6	**79 Eurynome**	107	6	100.0	[5] 5.6
3973 1981UC1[1]	6	27 Euterpe	33	6	100.0	[6] 18.2
5300[1]	6	–	–	–	–	–
2914 1965SB[1]	5	–	–	–	–	–
4097 Tsurugisan[1]	5	–	–	–	–	–
–	–	*1978 PO3*[2]	6	–	–	–

(*) Corresponding to a WAM skeleton found at the smallest scale only.

Remarks :
[1] HCM clump
[2] WAM marginal grouping
[3] 2 objects in WAM's 20
[4] $\%_{WAM} = 41.2$ (**)
[5] $\%_{WAM} = 30.0$ (**)
[6] $\%_{WAM} = 42.9$ (**)
(**) Considering the corresponding WAM subcluster only.

Table 4. Same as Table 3 in the outer zone

Id(HCM)	N_{HCM}	Id(WAM)	N_{WAM}	N_{com}	$\%_{HCM}$	$\%_{WAM}$
24 Themis	550	24 Themis	517	491	89.3	95.0
221 Eos	477	**221 Eos**	482	444	93.1	92.1
158 Koronis	325	**158 Koronis**	299	289	88.9	96.7
10 Hygiea	103	10 Hygiea	175	97	94.2	(4) 55.4
490 Veritas	22	92 Undina	36	22	100.0	61.1
137 Meliboea	13	137 Meliboea	16	13	100.0	81.3
1298 Nocturna	18	10 Hygiea	175	18	100.0	(5) 10.3
3330 1985RU1 (1)	14	(*)	–	–	–	–
293 Brasilia (1)	10	**293 Brasilia**	18	10	100.0	(6) 55.6
1070 Tunica (1)	10	(*)	–	–	–	–
1041 Asta (1)	7	(*)	–	–	–	–
1981EO19 (1)	7	–	–	–	–	–
778 Theobalda (1)	6	(*)	–	–	–	–
845 Naëma (1)	6	**845 Naëma**	7	6	100.0	85.7
507 Laodica (1)	5	(*)	–	–	–	–
366 Vincentina (1)	5	(*)	–	–	–	–
– (3)	–	283 Emma	15	–	–	–
–	–	918 *Itha* (2)	9	–	–	–
–	–	627 *Charis* (2)	7	–	–	–

(*) Corresponding to a not significant WAM grouping.

Remarks :
(1) HCM clump
(2) WAM marginal grouping
(3) Eos' subgrouping for HCM
(4) $\%_{HCM} = 83.5$, $\%_{WAM} = 94.5$ (**)
(5) $\%_{HCM} = 94.4$; $\%_{WAM} = 73.9$ (**)
(6) $\%_{WAM} = 100.0$ (**)
(**) Considering the corresponding WAM subcluster only.

Table 5. Same as Tables 3 and 4 in the medium zone

Id(HCM)	N_{HCM}	Id(WAM)	N_{WAM}	N_{com}	$\%_{HCM}$	$\%_{WAM}$
15 Eunomia	439	**15 Eunomia**	303	298	67.9	98.3
1 Ceres	89	**93 Minerva**	88	84	94.4	95.5
170 Maria	77	**170 Maria**	83	74	96.1	89.2
668 Dora	77	**668 Dora**	79	75	97.4	94.9
145 Adeona	63	**145 Adeona**	67	61	96.8	91.0
125 Liberatrix	44	847 Agnia[3]	74	35	79.5	47.3
110 Lydia	26	110 Lydia[5]	50	26	100.0	52.0
808 Merxia	26	**808 Merxia**	29	26	100.0	89.7
569 Misa	25	**569 Misa**	27	23	92.0	85.2
1726 Hoffmeister	22	110 Lydia[4]	50	15	68.2	[6] 30.0
2085 Henan	22	847 Agnia	74	19	86.4	25.7
410 Chloris	21	**410 Chloris**	27	21	100.0	77.8
1644 Rafita	21	**1644 Rafita**	23	21	100.0	91.3
128 Nemesis	20	58 Concordia	38	20	100.0	[7] 52.6
1128 Astrid	10	**1128 Astrid**	11	10	100.0	90.9
1639 Bower	10	**342 Endymion**	15	10	100.0	66.7
2514 Taiyuan	9	847 Agnia	74	9	100.0	[8] 12.2
3 Juno[1]	8	**3 Juno**	18	8	100.0	44.4
46 Hestia[1]	8	**46 Hestia**	13	8	100.0	61.5
751 Faina[1]	8	**170 Maria**	83	4	50.0	[9] 4.8
2299 Hanko[1]	8	2299 Hanko	8	6	75.0	75.0
3142 Kilopi[1]	7	–	–		–	–
1982QG[1]	7	(*)	–	–	–	–
144 Vibilia[1]	6	144 Vibilia	11	6	100.0	54.5
396 Aeolia[1]	6	396 Aeolia	12	6	100.0	50.0
1547 Nele[1]	6	1547 Nele	6	4	66.7	66.7
2198 Ceplecha[1]	6	–	–	–	–	–

Table 5. (continued)

Id(HCM)	N_{HCM}	Id(WAM)	N_{WAM}	N_{com}	$\%_{HCM}$	$\%_{WAM}$
157 Dejanira[(1)]	5	157 Dejanira	13	5	100.0	[(6)] 38.5
4945[(1)]	5	(*)	–	–	–	–
1981EG1[(1)]	5	124 Alkeste	8	1	20.0	12.5
(*)	–	**4700 Carusi**	8	–	–	–
(*)	–	**237 Coelestina**	6	–	–	–
–	–	**779 Nina**	20	–	–	–
–	–	**355 Gabriella**	14	–	–	–
–	–	**1121 Natascha**	11	–	–	–
–	–	**3039 Yangel**	10	–	–	–
–	–	**140 Irma**	7	–	–	–
–	–	**173 Ino**	7	–	–	–
–	–	649 Josepha	18	–	–	–
(*)	–	1772 Gagarin	18	–	–	–
–	–	272 Antonia	12	–	–	–
–	–	1294 Antwerpi	11	–	–	–
–	–	5153	10	–	–	–
–	–	3393 Stur	8	–	–	–
–	–	1460 Tatun	7	–	–	–
–	–	*262 Valda*[(2)]	38	–	–	–
(*)	–	*55 Pandora*[(2)]	27	–	–	–
–	–	*421 Zahringia*[(2)]	24	–	–	–
(*)	–	*1292 Luce*[(2)]	21	–	–	–
(*)	–	*322 Phaeo*[(2)]	14	–	–	–
–	–	*269 Tirza*[(2)]	10	–	–	–

(*) Corresponding to a not significant HCM, or WAM, grouping.

Table 5. (end)

Remarks :
(1) HCM clump
(2) WAM marginal grouping
(3) 3 objects in WAM's 58
(4) 7 objects in WAM's 272
(5) $\%_{WAM} = 92.9$ (**)
(6) $\%_{WAM} = 100.0$ (**)
(7) $\%_{HCM} = \%_{WAM} = 80.0$ (**)
(8) $\%_{WAM} = 90.0$ (**)
(9) $\%_{WAM} = 66.7$ (**)
(**) Considering the corresponding WAM subcluster only.

6 Conclusions

One conclusion of this review on what has been done and what we know on the identification of dynamical families is that we are at the beginning of a new era in the field of asteroid families. The reason is that physical studies of families can be now undertaken on the basis of a very solid statistical evidence about the groupings that can be recognized in the asteroid belt today. Of course, we cannot state that we know all the existing families, but we believe that those that are recognizable today on the basis of the available observational data have been found, and that the possibility to consider as real families some random groupings is fully negligible, at least in the cases of the most robust families listed in the last Section. These candidates (clans or clusters) are statistically significant, have passed several robustness tests and have been found by two independent methods. Therefore, the main problem now seems to be the scarcity of available physical data for most of the family members. This problem will be overcome in the near future by the several observational campaigns, mainly in the field of spectroscopy, that are presently carried out or planned. Some preliminary results of this effort have been the spectacular results concerning the Vesta family. We expect to have other similar cases in a few years. A special attention should be devoted in particular to the members of the clans of Nysa and Hygiea, since the former could be the main source of the uncommon F-type asteroids in the inner belt, while the latter can be expected to be another example of an energetic cratering event like that of Vesta.

Beside the taxonomic determination of the family members, diameter estimates of the family members are of the highest importance. Cellino et al. (1991) have shown that the diameter distribution of family members has different characteristics with respect to the distribution of the non family member asteroids. Moreover the analysis of mass distribution within family members can provide some essential information on the physics of the collisional events responsible of the family formations, and possibly about the ages of the different families, on the basis of the overall slope of the size distribution, that could be time-dependent according to a recent analysis by Marzari et al. (1994). Some clues about the age of families can also be reached in some cases on the basis of the dynamical evidence. A good example is given by the family of Veritas, for which it has been shown that the proper elements of some of its members are not stable over short time scales, what can have deep implications about the age of this family (Milani and Farinella 1994).

Another interesting topic is related to the dust bands discovered by the IRAS satellite. These dust bands can be associated to the main asteroid families (Sykes et al. 1989; Dermott et al. 1989).

Finally, we can remind that the first direct imaging of a family member has been shocking. The Galileo images of 243 Ida, a member of the big Koronis family, have shown the presence of a small satellite, in addition to an unexpectedly high cratering of the asteroid surface. This is only another example showing that the physical investigation of asteroid families, which is now possible on the basis of their firm dentification, is one of the most interesting and challenging fields of planetary research for the next years.

References

Bendjoya, Ph. (1993): A classification of 6479 asteroids into families by means of the wavelet clustering method. A&A Suppl. Ser. **102**, 25–55.

Bendjoya, Ph., Cellino, A. (1992): Asteroid families identified by two different methods. In Benest D., Froeschlé C. (Eds.) *Interrelations between physics and dynamics for minor bodies in the Solar System*, Editions Frontieres, pp. 19–43.

Bendjoya, Ph., Cellino, A., Froeschlé, C.,Zappalà, V. (1993): Asteroid dynamical families : a reliability test for new identification methods. A&A **272**, 651–670.

Bendjoya, Ph., Slézak, E., Froeschlé, C. (1991): The wavelet transform : a new tool for asteroid family determination. A&A **251**, 312–330.

Benz, W., Asphaugh, E. (1994): Impact simulations with Fracture : I. Method and Tests. Icarus **107**, 98–116.

Benz, W., Asphaugh, E., Ryan, E.V. (1994): Numerical Simulations of Catastrophic Disruption : Recent Results. Planet. and Space Sci. **42**, 1053–1066.

Binzel, R.P., Bus, S.J., Xu, S. (1994): Physical studies of small asteroid families. B.A.A.S **26**, 1178.

Binzel, R.P., Xu, S. (1993): Chips of asteroid Vesta : evidence for the parent body of basaltic achondrite meteorites. Science **260**, 186–191.

Brouwer, D. (1951): Secular variations of the orbital elements of the minor planets. Astron. J. **56**, 9–32.

Bottke, W.F., Nolan, M.C., Greenberg, R., Kolvoord, R.A. (1994): Velocity distribution among colliding asteroids. Icarus **107**, 255–268.

Cellino, A., Zappala, V., Farinella, P. (1991): The size distribution of main belt asteroids from IRAS data. Mon. Not. R.A.S. **253**, 561–574.

Chapman, C.R., Paolicchi, P., Zappala, V., Binzel, R.P., Bell, J.F. (1989): Asteroid families: physical properties and evolution. In Binzel R.P., Gehrels T., Matthews M.S. (Eds.) *Asteroids II*, Univ. of Arizona Press, pp. 386–415.

Davis, D.R., Ryan, E.V. (1990): On collisional disruption : Experimental results and scaling laws. Icarus **83**, 156–182.

Dermott, S.F., Durda, D.D., Gustafson, B.A.S., Jayaraman, S., Liou, J.C., Xu, Y.L. (1994): Zodiacal Dust Bands. In Milani A., Di Martino M., Cellino A. (Eds.) *Asteroids, Comets, Meteors 1993*, Kluwer, pp. 127–142.

Farinella, P., Davis, D.R., Cellino, A., Zappala, V. (1992): From asteroid clusters to families : A proposal for a new nomenclature. In Harris A.W., Bowell E. (Eds.) *Asteroids, Comets, Meteors 1991*, Lunar and Planetary Institute, Houston, pp. 165–166.

Fujiwara, A., Cerroni, P., Davis, D.R., Ryan, E., Di Martino, M., Holsapple, K., Housen, K. (1989): Experiments and scaling laws on catastrophic collisions. In Binzel R.P., Gehrels T., Matthews M.S. (Eds.) *Asteroids II*, Univ. of Arizona Press, pp. 240–265.

Giblin, I., Martelli, G., Smith, P.N., Cellino, A., Di Martino, M., Zappala, V., Farinella, P., Paolicchi, P. (1994): Field fragmentation of macroscopic targets simulating asteroidal catastrophic collisions. Icarus **110**, 203–224.

Hirayama, K. (1918): Groups of asteroids probably of common origin. A.J. **31**, 185–188.

Knežević, Z., Milani, A. (1994): Asteroid proper elements : The big picture. In Milani A., Di Martino M., Cellino A. (Eds.) *Asteroids, Comets, Meteors 1993*, Kluwer, pp. 143–158.

Melosh, H.J., Ryan, E.V., Asphaug, E. (1992): Dynamic Fragmentation in Impacts : Hydrocode Simulation of Laboratory Impacts. Jour. Geoph. Res. **97/E9**, 14735–14759.

Milani, A., Farinella, P. (1994): The age of Veritas deduced from chaotic chronology. Nature **370**, 40–42.

Milani, A., Knežević Z (1992): Asteroid proper elements and secular resonances. Icarus **98**, 211–232.

Milani, A., Knežević, Z. (1994): Asteroid proper elements and the dynamical structure of the asteroid belt. Icarus **107**, 219–254.

Morbidelli, A., Zappalà, V., Moons, M., Cellino, A., Gonczi, R. (1995): Asteroids families close to mean motion resonances. Dynamical effects and physical implications. Icarus **118**, 132–154.

Muinonen, K., Bowell, E. (1993): Asteroid orbit determination using Bayesian probabilities. Icarus **104**, 255–279.

Murtagh, F., Heck, A. (1987): *Multivariate data analysis*, Reidel.

Nakamura, A., Suguiyama, K., Fujiwara, A. (1992): Velocity and spin of fragments from impact disruptions : an experimental approach to a general law between mass and velocity. Icarus **100**, 127–135.

Paolicchi, P., Cellino, A., Farinella, P., Zappalà, V. (1989): A semiempirical model of catastrophic breakup processes. Icarus **77**, 187–212.

Paolicchi, P., Verlicchi, A., Cellino, A. (1996): An improved semi-empirical model of catastrophic impact processes I. Theory and laboratory experiments. Icarus **121**, 126–157.

Scholl, H., Froeschlé, Ch., Kinoshita, H., Yoshikawa, M., Williams, J.G (1989): In Binzel R.P., Gehrels T., Matthews M.S. (Eds.) *Asteroids II*, Univ. of Arizona Press, pp. 845–862.

Sykes, M.V., Greenger, R., Dermott, S.F., Nicholson, P.D., Burns, J.A., Gautier, T.N. (1989): In Binzel R.P., Gehrels T., Matthews M.S. (Eds.) *Asteroids II*, Univ. of Arizona Press, pp. 336–367.

Williams, J.G. (1979): Proper orbital elements and family memberships of the asteroids. In Gehrels T. (Ed.) *Asteroids*, Univ. of Arizona Press, pp. 1040–1063.

Williams, J.G. (1989): Asteroid family identification and proper elements. In Binzel R.P., Gehrels T., Matthews M.S. (Eds.) *Asteroids II*, Univ. of Arizona Press, pp. 1034–1072.

Zappalà, V., Bendjoya, P., Cellino, A., Farinella, P., Froeschlé, C. (1995): Asteroid families: Search of a 12,487-asteroid sample using two different clustering techniques. Icarus **116**, 291–314.

Zappalà, V., Cellino, A. (1992): Asteroid Families : Recent Results and Present Scenario. Celest. Mech. **54**, 207–227.

Zappalà V., Cellino, A., Farinella, P. (1992): 'A comparison between families obtained from different proper elements. in *Asteroids, Comets, Meteors 1991*, A.W. Harris & E. Bowell (Eds.), Lunar and Planetary Institute, Houston, pp. 675–678.

Zappalà, V., Cellino, A., Farinella, P., Knežević, Z. (1990): Asteroid families : identification by hierarchical clustering and reliability assessment. A.J. **100**, 2030–2046.

Zappalà, V., Cellino, A., Farinella, P., Milani, A. (1994): Asteroid Families : Extension to Unnumbered Multi-opposition Asteroids. A.J. **107**, 772–801.

Origin and Dynamical Transport of Near-Earth Asteroids and Meteorites

Alessandro Morbidelli and Christiane Froeschlé

O.C.A. Observatoire de Nice, B.P. 4229, F-06304 Nice Cedex 4, France

Origine et transport dynamique des NEAs et des météorites

Résumé. Au début du siècle, les scientifiques croyaient que les météorites étaient des objets venant de l'espace interstellaire et évoluant sur des orbites hyperboliques. Actuellement on sait que la plus grande partie de ces petits corps sont des fragments d'astéroïdes de la Ceinture Principale. Cependant, jusqu'à ces dernières années, on ne savait pas comment ces fragments pouvaient atteindre la surface de notre planète. Dans ce chapitre, on décrira les principaux mécanismes de transport qui permettent à ces fragments d'astéroïdes d'arriver sur la surface terrestre, et qui sont alors appelés météorites. Nous verrons que ces mécanismes sont tous associés aux phénomènes de résonances. Nous concentrerons notre exposé sur la description phénoménologique de ces mécanismes, en évitant les aspects mathématiques.

Abstract. This chapter describes the most important mechanisms for transporting material from the Asteroid Belt to the Earth's vicinity and the evolution of bodies on Earth-crossing orbit. Particular care will be paid to evolutionary timescales and statistical aspects of meteoroids dynamics.

1 Introduction

At the beginning of this century it was believed that meteorites come from the interstellar matter travelling on hyperbolic orbits. Today the birth place of most meteorites and NEAs (Near-Earth-asteroids) has been recognized to lie in the main asteroid belt. However, until recently, serious problems remained in understanding the asteroid-meteorite connexion, i.e. the mechanisms of transport. Indeed fragments resulting from collisions between asteroids are generally ejected at velocities of the order of 100 m/s, far too small for directly causing the drastic orbital changes needed to achieve planet-crossing orbits. On the other hand impulsive velocity increments of several km/s would induce shocks and thermal modifications in the meteoritic material, which are not generally observed. Thus the interplay between collisions and subtle dynamical effects is required to transport material from the asteroid belt to the Earth. A long-standing issue in planetary science is that of identifying and estimating the efficiency of the dynamical routes. In recent years significant progress has been achieved on these problems and in this paper we will give a description of the main different dynamical transport mechanisms which all involve resonances.

There are two principal classes of resonances which affect the motion of small bodies in the Solar System. These are the mean motion resonances (resonances among the periods on the orbits) and the secular resonances (resonances among the precession rates of the orbits).

As a matter of fact, when one considers the distribution of the asteroids in a diagram semi major axis a versus the inclination I (Fig. 1), the resonant regions are mostly empty of asteroids, which means that the resonances are efficient mechanisms to eject, more or less rapidly, asteroids located in their regions.

In the following we will describe in more details the mechanisms of transport related to the resonances mentioned above. We will not enter in details concerning Celestial Mechanics. We prefer rather to take the phenomenological point of view, so that we will start from numerical integrations to describe the main features of these mechanisms (regular or/and chaotic behaviour, strength, time scales, etc.). We start in Sect. 2 by recalling the different elements of an orbit, and show which are the problematics of the transport mechanism. Section 3 is devoted to secular resonances, with a particular care to the ν_6 one. Section 4 will be devoted to mean motion resonances. Then, Sect. 5 will describe how asteroids on planet crossing orbits evolve under the influence of planetary close encounters. Finally Sect. 6 will describe the timescales of the Near Earth Asteroids (NEAs) evolution and their expected orbital distribution, while Sect. 7 will discuss achieved results and open problems.

2 Geometry of Orbits

The elliptic orbit of any body moving around the Sun is characterized by six osculating elements (Fig. 2).

- the semi major axis a, which is the length of CA.

- the eccentricity e, which is the ration CF_1/CA.

- the inclination I of the orbital plane with respect to the reference plane.

- the longitude of the ascending node Ω, which is the angle between the reference direction (usually the γ point) and the intersection point of the orbital plane and the reference plane.

- the perihelion argument ω, which is the angular distance between the ascending node and the pericenter direction. The perihelion is the point of the orbit, closest to the Sun, *the perihelion distance* $p = a(1 - e)$.

- the true anomaly f which is the angular distance of the body with respect to the pericenter direction.

Notice that the two first elements a and e determine the dimension and the form of the orbit, the value of e discriminates among ellipses ($e < 1$), parabolae ($e = 1$) and hyperbolae ($e > 1$). When $e = 0$ the ellipses degenerate into circular orbits. I and Ω determine the orbital plane, ω the orientation of the orbit in the plane, and f the position of the object on its orbit.

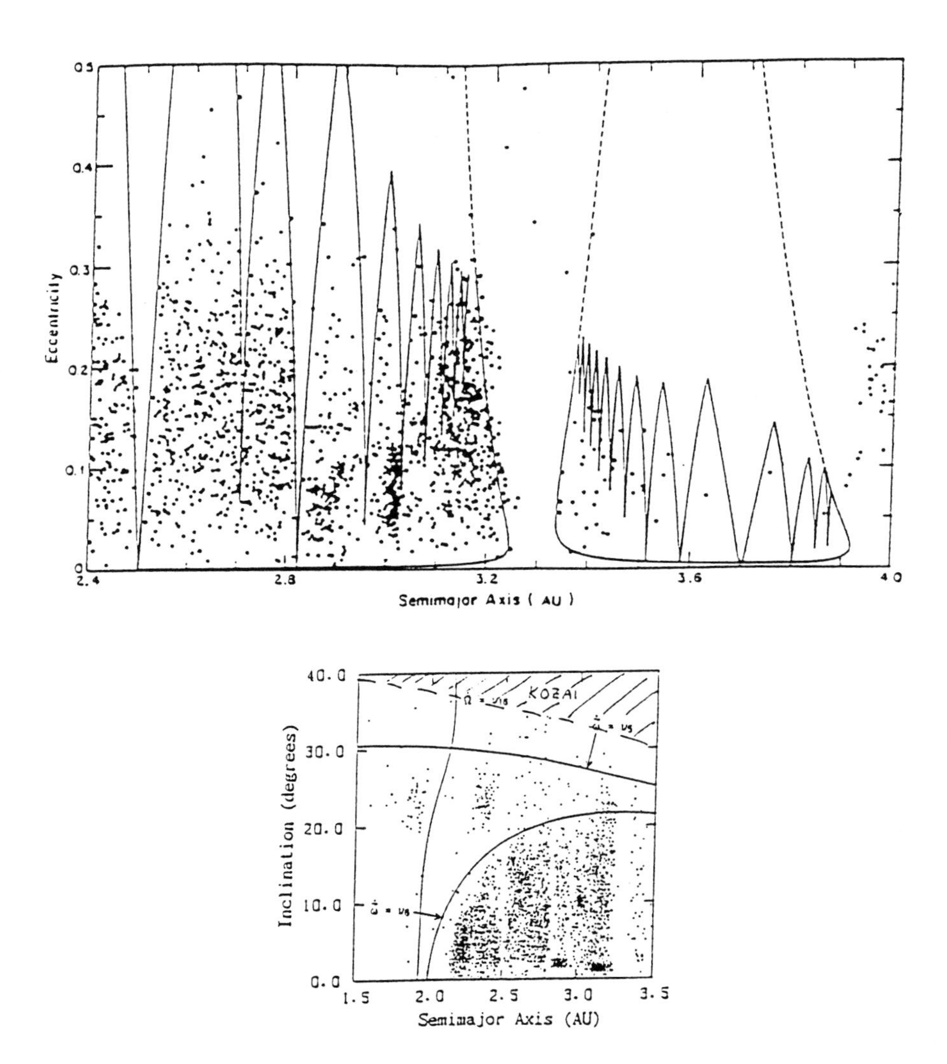

Fig. 1. The location of numbered asteroids in the $a - e - I$ space. The picture above (from Dermott and Murray 1993) points out the gaps corresponding to mean motion resonances the borders of which are V-shaped on the $a - e$ plane. The picture below (from Williams 1969), shows the location of the main secular resonances; it is evident that secular resonances play a role in the confinement of the distribution of the asteroids in the $a - I$ plane. The dashed region at large inclination, depleted of asteroids, corresponds to the region dominated by the Kozai resonance, a particular secular resonance that occurs when the precession rate of the asteroid's longitudes of perihelion is equal to that of the node.

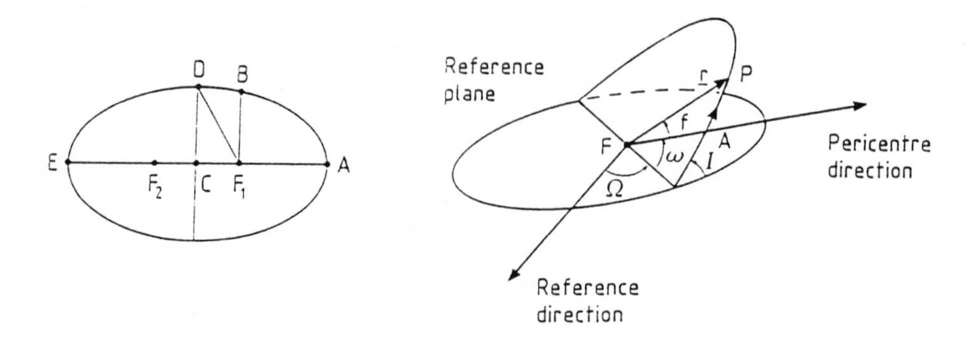

Fig. 2. On the left, the shape of an elliptic orbit defines the semimajor axis a and the eccentricity e; F_1 and F_2 are the two focal points. On the right, the orientation of the orbit in the space defines the inclination I, the longitude of node Ω and the argument of perihelion ω. The true anomaly f denotes the current position of the object on the orbit.

The longitude of perihelion is equal to: $\varpi = \omega + \Omega$

On Fig. 3a are drawn three orbits having the same value of the semi major axis and different values of the eccentricity, (e = 0., 0.5, and 0.8): the ellipse is more and more elongated with increasing value of e. The Sun is always in the *focus* of the ellipse, so that, increasing the eccentricity, the perihelion of the orbit approaches the Sun, while the aphelion's distance increases. So, in order to transport objects from one region to another in the Solar System, the most efficient way is to pump the eccentricity; for example, as shown on Fig. 3b, an asteroid orbiting around the Sun on a circular orbit of semi major axis $a = 3$ AU (the AU is the unit length equal to the mean distance Earth-Sun, i.e. approximatively 150 millions of kilometers), becomes Earth-crosser, and also approaches Jupiter ($a_J = 5.2$ AU), if its eccentricity is increased up to 0.66.

How can the eccentricity of an orbit change so much? As we describe in the following, the two types of resonances mentioned before are the most efficient ones to pump the eccentricity of orbits.

3 The Secular Resonances

The planets give rise to mutual perturbations, consequently their orbits are affected by very slow (i.e. secular) precessions, with periods ranging from thousands to millions years. The precession of the perihelion is the slow motion of rotation of the orbit in its plane, and the precession of the node is the rotation of the orbit in the space. For simplicity, if one restrict to the Sun-Jupiter-Saturn system these motions are quasi-periodic with three basic frequencies:

- g_5 which is the average precession rate of Jupiter's longitude of perihelion $\varpi_J = \omega_J + \Omega_J$

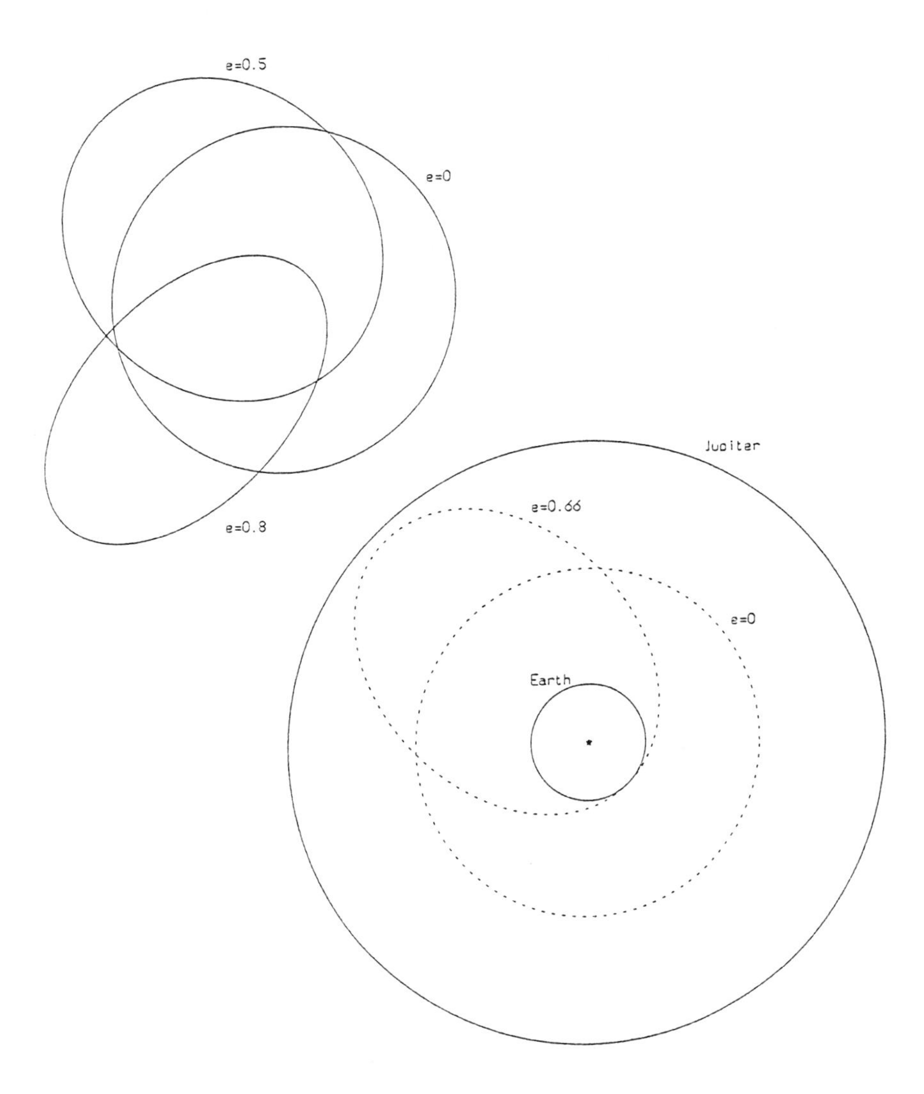

Fig. 3. a (Top left) the shape of three orbits with the same semimajor axis and different values of the eccentricity e; the ellipses are more and more elongated, and the perihelion approaches the Sun which is always in the focus of the orbits. **b** (Bottom right) two orbits of asteroids with semimajor axis $a = 3$ AU are plotted with respect to the orbit of the Earth and of Jupiter; the one with $e = 0.66$ is tangent to the Earth and also approaches Jupiter. Then the increase of the eccentricity is a mechanism for transporting asteroids towards the planets.

- g_6 which is the average precession rate of Saturn's longitude of perihelion of Saturn $\varpi_S = \omega_S + \Omega_S$.

- s_6 which is the precession rate of both nodes.

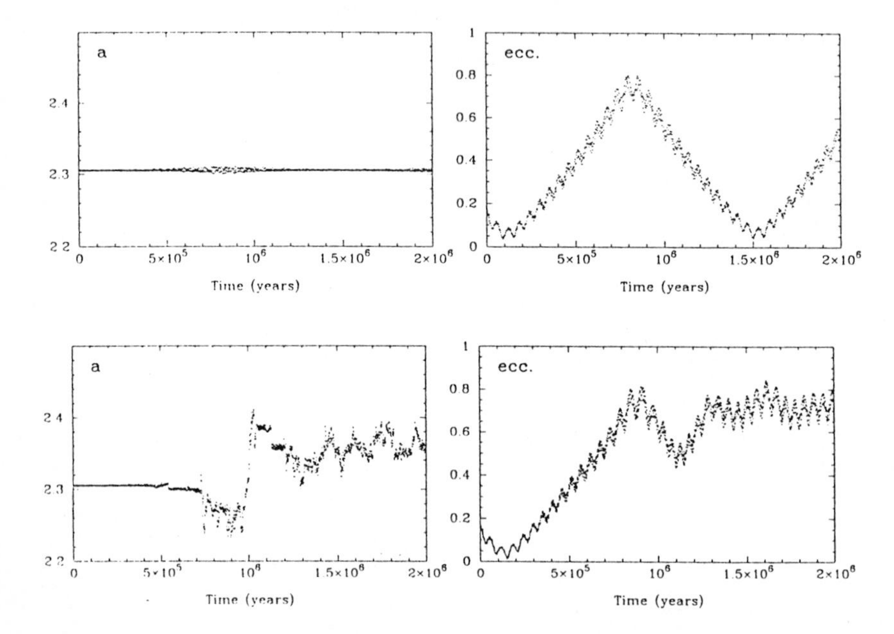

Fig. 4. a (Top) the evolution of the semimajor axis and of the eccentricity of a fictitious body in the ν_6 resonance. The effect of the inner planets are neglected. The eccentricity is pumped up and down in an extremely periodic and regular way. The amplitude of the eccentricity variation is very large. **b** (Bottom) the evolution of the same orbit taking into account the effect of the inner planets. As a consequence of the close encounters which are effective as soon as $e > 0.4$ (and especially $e > 0.6$), the semimajor axis is no longer constant but exhibits a sort of random walk. Then, the body is displaced with respect to the resonance and therefore the eccentricity does not decrease again to zero. Thus, in a realistic model where all planets are taken into account, the ν_6 resonance pumps up the eccentricity in a non reversible manner.

The planets exert also secular perturbations on any small body orbiting around the Sun and force the precession and the deformation of its orbit. We denote by g the precession rate of the asteroid's longitude perihelion, and by s the precession frequency of its node. These secular perturbations give particularly large effects when a *secular resonance* occurs, namely when the frequency of precession of the small body g or s becomes *nearly equal* to an eigenfrequency of the planetary system, namely when $g = g_5$, $g = g_6$ or $s = s_6$; these resonances are often called respectively ν_5, ν_6 and ν_{16}.

Though the phenomena of secular resonances have been known since the end of the last century, the first modern approach to study secular resonances in a theoretical way was done by Williams (1969). The first numerical works and quantitative results on the evolution of resonant bodies appeared only in the last decade (see Froeschlé and Scholl 1989 for a review) with the development of high speed computers, since the timescales on which secular resonances act are very long, of the order of 1 million of years. These numerical works revived the interest for the study of secular resonances and as a consequence new improved theories appeared (see Froeschlé and Morbidelli 1993 for a review).

Among the three main secular resonances, the ν_6 one provides the most effective mechanism as far as material transport is concerned. As a matter of fact, the ν_6 resonance is particularly strong, at least one order of magnitude stronger than one could reasonably expect, since it can pump up the eccentricity from 0 to almost unity. Conversely, the secular resonance ν_5 increases the eccentricity only to values which result in Mars grazing or Mars crosser orbits, while the secular resonance ν_{16} involving the precession of the node may produce only large increase in inclination and no dramatic changes in the eccentricity. Moreover, while the ν_6 resonance bounds the inner asteroid belt, a very densely populated region, the ν_5 and ν_{16} resonances cross regions where the asteroid population density is very low (Fig. 1).

In Fig. 4a we show an example of the evolution of the orbit of a (fictitious) object in the ν_6 resonance. The orbit has been integrated in the Sun-body-Jupiter-Saturn system. The semimajor axis is constant along the full evolution over 2 Myr, while the eccentricity is pumped up and down in a very regular manner, oscillating from almost 0 to 0.8.

In Fig. 4b, we show the evolution of the same orbit, taking into account also the role of Mars and the Earth. The first part of the dynamical history is the same as in Fig. 4a. Then, since $e > 0.4$, close encounters with Mars start to happen. This produces small random changes in the semimajor axis. When $e > 0.6$, the orbit can have close approaches to the Earth. This gives more relevant random changes of a, which exhibits a sort of random-walk (see Sect. 5 for more details). As a consequence, the body is displaced with respect to the resonance. The eccentricity, therefore, does not decrease to 0 again, and in the second part of the integration oscillates between 0.6 and 0.8, which is a typical meteorite value. Therefore, we can conclude that the ν_6 resonance pumps the eccentricity in a irreversible manner, when the effect of the inner planets is taken into account.

Since the ν_6 resonance can produce planet-crossing orbits by eccentricity pumping, one can expect to find a fraction of the present population of planet crosser objects inside the ν_6 resonance or very close to it. This has been checked by Froeschlé et al. (1995), who found out that 19 of the considered sample of 181 objects are presently related to the resonance. This result is illustrated in Fig. 5 where the position of the 181 asteroids is projected onto the frequency plane $g - s$ in order to point out the role of secular resonances. The crosses denote the proper frequencies of the objects which are not related to some resonance. The circles, conversely, denote the approximate frequencies of those bodies which are very close to one resonance. The two vertical lines denote the ν_5 secular resonance at $g = 4.257$"/yr and the ν_6 secular resonance at $g = 28.245$"/y; the horizontal line denotes the ν_{16} resonance at $s = -26.345$"/yr.

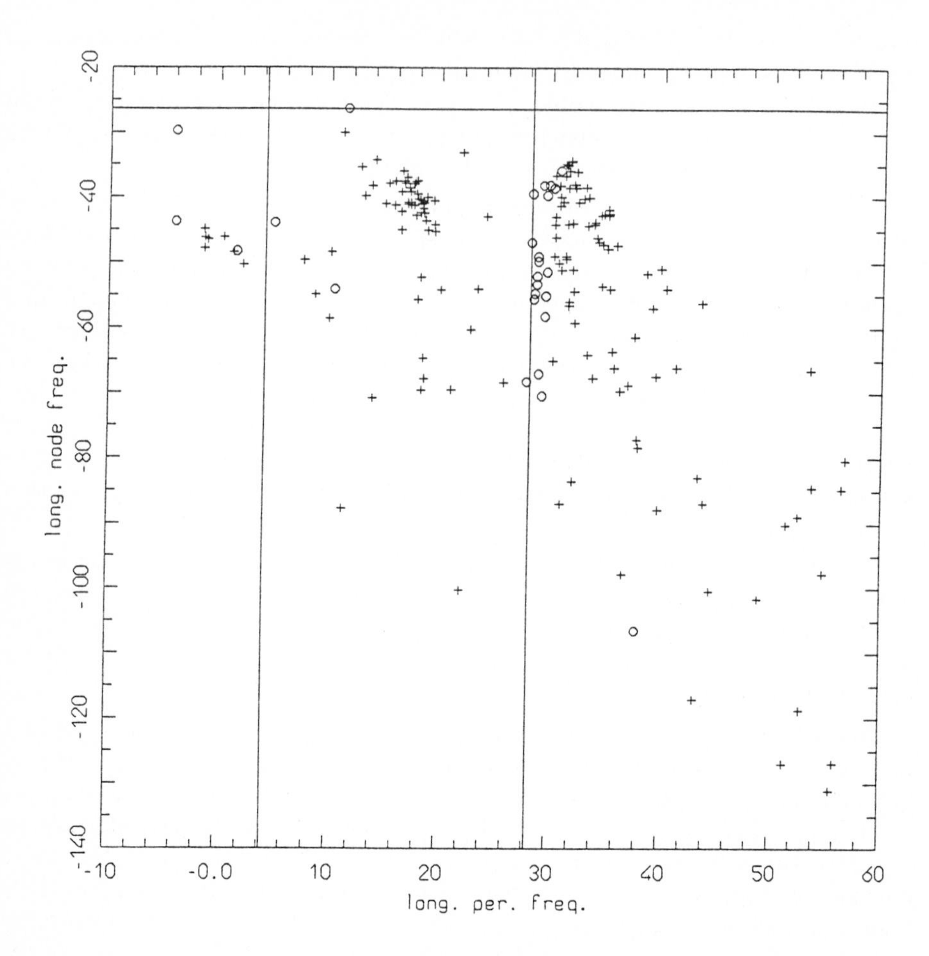

Fig. 5. The secular frequencies g and s of 181 NEAs. Dots represent objects which are inside some resonance or close by. Crosses denote asteroids which are far from the main resonances. The two vertical lines denote the ν_5 and ν_6 resonances; the horizontal line denotes the ν_{16} resonance. The dots which are not on these lines are connected to some other resonance, for example a mean motion one (see Sect. 4).

Froeschlé et al. (1995) integrated numerically over some millions of years the 19 bodies which are close to the ν_6 resonance, and found that 16 of them have minimal eccentricities smaller than 0.2, which is a typical value of the main belt population. This confirms the conjecture that these bodies are fragments of main belt asteroids fallen in the ν_6 transport route.

Farinella et al. (1994) first pointed out that the ν_6 resonance is very efficient in pumping the eccentricity to values very close to unity, so that the perihelion distance becomes smaller than the Sun's radius, forcing the resonant body to collide with our star. In fact, 5 of the 19 bodies in Froeschlé et al. survey fall into the Sun: an example of that is given by Fig. 6.

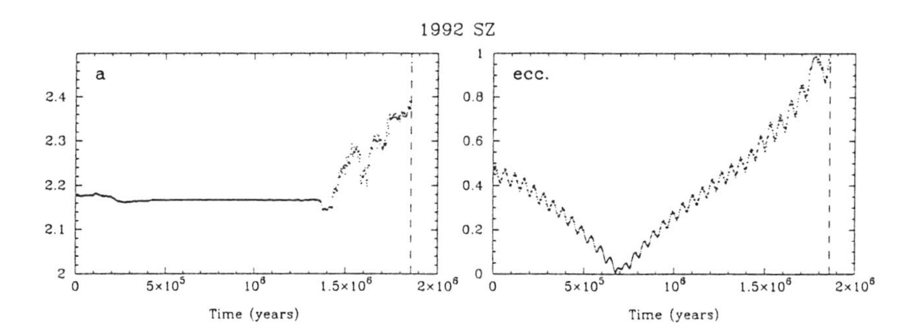

Fig. 6. the evolution of the NEA 1992 SZ located into the secular resonance ν_6: the eccentricity is pumped up so much that the asteroid finally falls into the Sun.

4 Mean Motion Resonances

A second class of resonances may also pump up the eccentricity of a small body: this is the class of the so-called mean motion resonances. We know that for any body orbiting around the Sun, the orbital period T and the semi major axis a verify the 3rd Kepler law namely $T^{-2}a^3 = Constant$, which means that the period of revolution of an object around the Sun depends only on the semimajor axis of the orbit and does not depend on the eccentricity, inclination and orientation of the orbit.

A mean motion resonance occurs when the orbital period of a minor body is commensurable with the Jupiter's orbital period. For example considering the 3/1 resonance, the minor body performs 3 revolutions around the Sun, while Jupiter performs 1 revolution. As a consequence, after the corresponding interval of time the same relative position of the minor body with respect to Jupiter is recovered. Then for these resonant orbits, where the geometrical configurations are repeated almost identically, the gravitational perturbations due to Jupiter are accumulated, and their effects may considerably modify the initial orbit.

The mean motion resonances have been studied since the beginning of Celestial Mechanics. Indeed they appear already in the most simple non-integrable problem, i.e. the restricted circular planar three body problem Sun-asteroid-Jupiter. Moreover, already in 1866 Kirkwood discovered that in correspondence of the main mean motion resonances, evident gaps in the population of main belt asteroids appear.

The restricted circular planar three body problem is a mathematical model where only the perturbations of one planet on a circular orbit are taken into account and the mass and the inclination of the minor body are neglected. Of course this model is not realistic for the description of the dynamics in the Solar System, since the eccentricity and the precession of planetary orbits are neglected. In the frame of this simple model, the mean motion resonances are not dangerous regions for the dynamical life of a minor body, neither they are

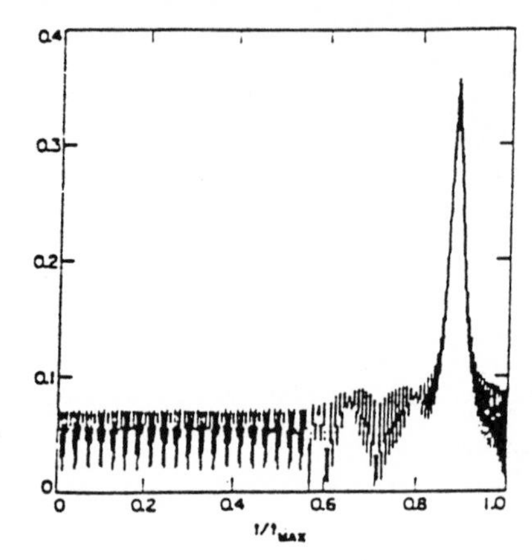

Fig. 7. Chaotic jumps of the eccentricity in the 3/1 mean motion resonance according to Wisdom (1983).

effective mechanisms of transport. Indeed, the only remarkable effect is the oscillation of the semi-major axis around the value corresponding to the exact resonance, while the eccentricity does not suffer very large changes.

The picture changes in a dramatic way when the eccentricity of Jupiter's orbit is taken into account, even if still neglecting the precession of Jupiter's orbit. In 1983, Wisdom pointed out the existence of a chaotic region in the 3/1 mean motion resonance which causes random jumps of the eccentricity up to 0.35 (Fig. 7). Wisdom pointed out that an asteroid located in the 3/1 resonance with an eccentricity equal to 0.35 is Mars grazer; then he conjectured that a close approach with Mars could change the semi-major axis of the asteroid, taking it away from the 3/1 resonance and even transporting the object to the Earth. This theory, on the one hand, provided for the first time a possible mechanism for the explanation of the origin of the 3/1 Kirkwood gap and, on the other hand, pointed out that a mean motion resonance can be a mechanism for transporting material (i.e. meteorites) to the inner planets. However, the time scales are very long, of order of 100 Myrs, since the probability to have an effective encounter with Mars is very small.

A similar approach, i.e. taking into account the eccentricity of Jupiter's orbit, has been followed by Yoshikawa (1989, 1990, 1991) also for what concerns other mean motion resonances; he also pointed out that in many cases (for example the 5/2 and 4/1 resonances) the eccentricity of the asteroid is pumped up to a Mars-crossing value. The behaviour of the eccentricity in the frame of this model is quite regular, like in the case of the ν_6 resonance discussed in Sect. 3.

The situation turns out to be substantially different according to realistic numerical integrations taking into account all the giant planets and therefore also the precession of their orbit. The behaviour of the orbit of the asteroid becomes strongly chaotic and also the maximal eccentricity reached is increased (Wisdom 1985, 1986; Yoshikawa 1989). More recently Morbidelli and Moons (1993) and Moons and Morbidelli (1994) pointed out that inside almost all mean motion resonances one can find the secular resonances ν_5, ν_6 and ν_{16}, and all these secular resonances overlap giving origin to a wide chaotic region. The chaotic region may extend at all eccentricities, so that the eccentricity e can jump up to Earth-crossing values (or even more, allowing collision with the Sun) in a very irregular manner and on a timescale of 1 Myr only. Close encounters with the Earth as well as Solar collisions explain easily the origin of the Kirkwood gaps (except that for the 2/1 mean motion resonance).

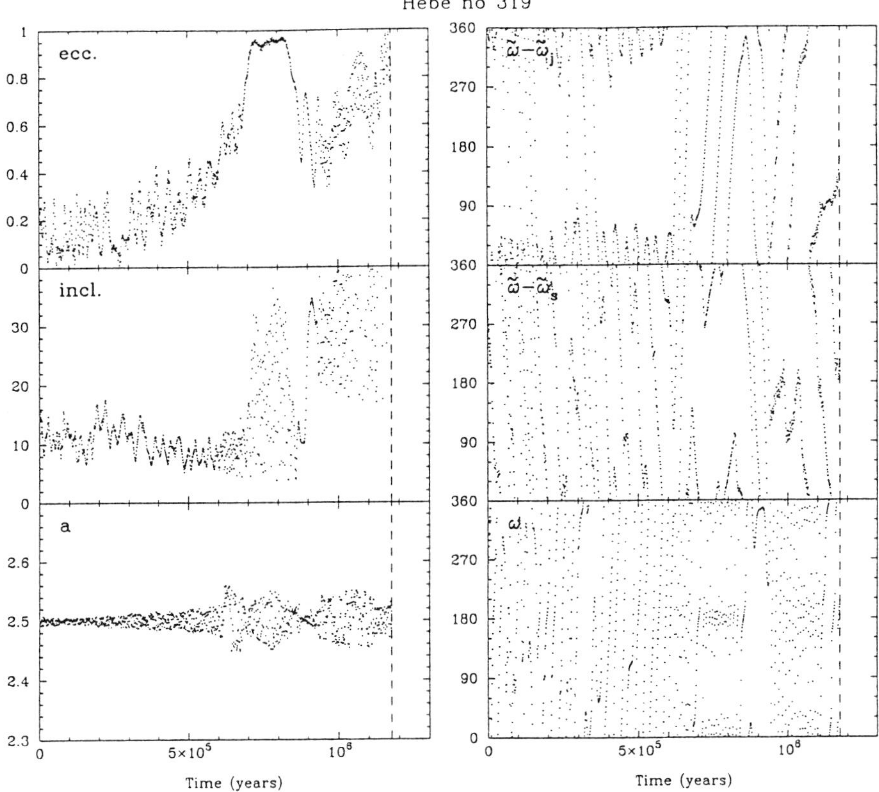

Fig. 8. The evolution of a fictitious object in the 3/1 mean motion resonance in a realistic numerical integration taking into account all the planets from the Earth to Saturn. The behaviour is completely chaotic and the eccentricity is pumped so much that the object falls into the Sun on a timescale of 1 Myr only.

As an example we reproduce in Fig. 8 a result from Farinella et al. (1993a) concerning a fictitious object in the 3/1 mean motion resonance; the integration takes into account the perturbations given by the planets from the Earth to Saturn. One sees immediately the strongly irregular behaviour of the eccentricity and of the inclination, in comparison to the results of Fig. 4. The semimajor axis oscillates widely around 2.5 AU which corresponds to the exact resonance. The amplitude of the oscillations of a changes also in an irregular way. This chaotic behaviour is precisely due to the temporary passage through the secular resonances ν_5 and ν_6 and through the so-called Kozai (1962) resonance. This can be checked by looking at the three critical angles of the three resonances, i.e. $\varpi - \varpi_J$, $\varpi - \varpi_S$ and ω respectively. The key for the understanding of these pictures is that any time there is an inversion in the motion of one of the critical angles, a resonance is crossed. The resonance crossings are correlated to the jumps in the eccentricity.

It has to be pointed out that this fictitious body is at the beginning of the integration time a typical main belt object since the eccentricity and the inclination are moderate ($e \simeq 0.15$ and $I \simeq 10°$). At the end, the eccentricity has increased so much that the object is Earth-crossing and falls into the Sun. This shows that the 3/1 Kirkwood gap can be formed in relatively short times and that the 3/1 resonance (like many other mean motion resonances) is an effective mechanism of transport of material to Earth-crossing orbit.

The analysis of the real NEAs points out that some of them (about 20 over 523) are presently in a mean motion resonance (mainly in the 3/1). It seems that objects in these mean motion resonances have a large probability to fall into the Sun (being the eccentricity pumped up to 1) in a few Myr.

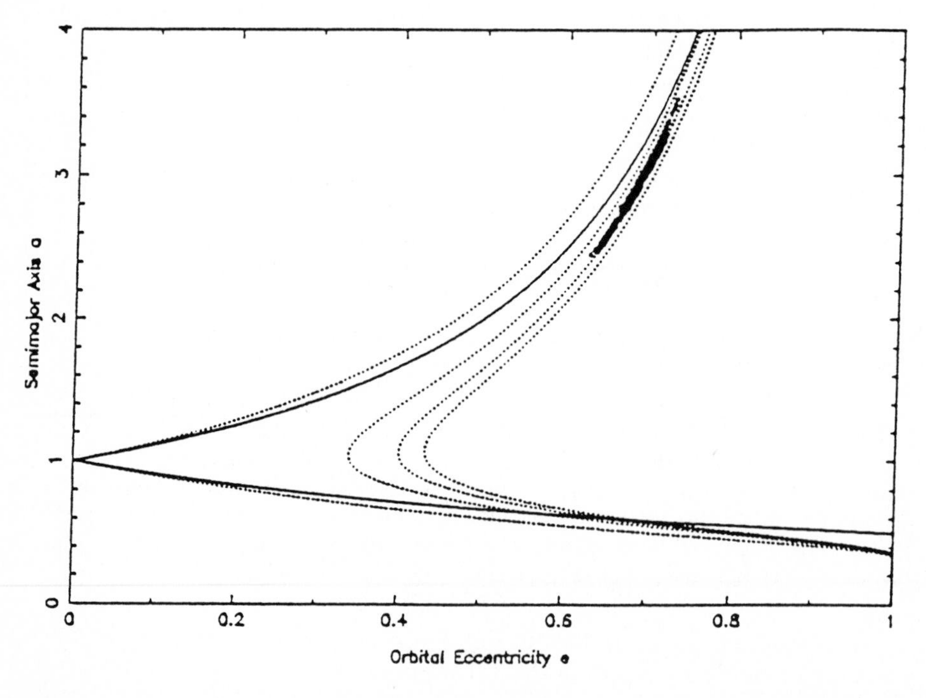

Fig. 9. The black dots show the evolution over 1 Myr of a test particle in the framework of the circular restricted three body problem Sun-Earth-asteroid. The two bold curves correspond to perihelion and aphelion distances equal to 1 AU. The dotted curves are contours of the Tisserand constant T with respect to the Earth (from Michel et al. 1996).

5 Transport of Planet Crossing Objects by Planetary Close Encounters

Objects evolving on orbits which intersect the orbit of a planet can encounter the planet from a close distance. Although small with respect to the Sun, the planet's gravity is the dominating force acting on an asteroid during a close encounter. As a result, the asteroid's heliocentric orbital elements may suffer a relevant impulse change. Such changes may easily be estimated in the case where the planet moves on a circular orbit, neglecting the Sun's influence during the encounter and following the classical Rutherford approach (Opik 1976). The asteroid's velocity vector $\mathbf{v}$ is changed by a quantity $d\mathbf{v}$ whose norm is

$$||d\mathbf{v}|| = \frac{U}{\sqrt{1+\left(\frac{U^2 b}{M}\right)^2}}$$

where U is the norm of the encounter velocity, b is the *impact parameter* which characterizes the encounter and M is the planet's mass. In the above formula the units of mass, space and velocity are the mass of the Sun, the planet's orbital radius and orbital velocity.

The velocity change $d\mathbf{v}$ results in a change of the asteroid's heliocentric orbital elements a, e and I of order $||d\mathbf{v}||$, its exact amount depending on the orientation of $d\mathbf{v}$ and on the asteroid's true anomaly at the moment of encounter. However, the changes in a e and I are correlated by the fact that U is unchanged during the encounter, so that the Tisserand parameter

$$T = 3 - U^2 = \frac{a'}{a} + 2\sqrt{\frac{a}{a'}(1-e^2)}\cos I \ ,$$

(where a' is the planet's semi major axis), is also unchanged. Figure 9 shows level curves of the Tisserand parameter with respect to the Earth on the plane (e, a) at $I = 0$, together with the evolution of a fictitious asteroid as it results from a numerical integration in the framework of the planar circular restricted three body problem Sun-Earth-asteroid. The asteroid is transported in the (e, a)-orbital space following a Tisserand parameter level curve, as expected in Opik's theory.

However, the situation changes dramatically if one takes into account a realistic model of the Solar System. The encountering planet's eccentricity forces the Tisserand parameter to change during the encounter, and, even more important, the secular perturbations due to planetary eccentricities, inclinations and precessions force the asteroid to evolve transversely to the Tisserand parameter level curves, mainly changing e at constant a (Fig. 10). As a result, the combination of planetary close encounters and secular perturbations allow the transport of the asteroid all over the planet crossing orbital space, spreading NEAs and meteoroids all over the (e, a) plane, even if all of them, when first become planet crossers, have specific values of semi major axis corresponding to the location of the transporting resonances.

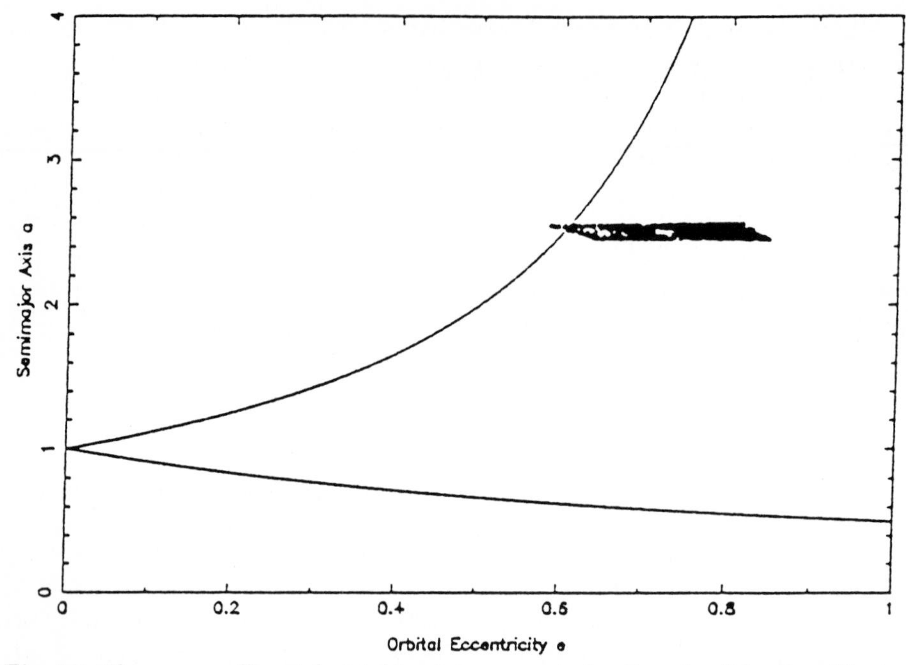

Fig. 10. The same as Fig. 9, but taking into account also the role of Jupiter. The test particle, being in the 3/1 resonance, has large oscillations in e at roughly constant a (from Michel et al. 1996).

6 Evolution Timescales and Orbital Distribution of NEAs

At the light of the results reported in the previous sections, Near Earth Asteroids and meteoroids should be the collisional fragments of the asteroid belt which have been injected into ν_6, 3/1 or other transporting resonances by the break-up event that gave them origin. They have rapidly evolved to Earth crossing orbits under the resonance effects and have been subsequently spread all over the Earth-crossing space by the combined action of planetary close encounters and secular perturbations. They orbit in the inner Solar System until they either collide with a planet, or hit the Sun or are ejected on hyperbolic orbit by a close encounter with Jupiter. It is generally expected that the number of bodies injected into resonance should balance the number of bodies eliminated by the above listed mechanisms, so to keep the NEAs population in a sort of steady state.

To quantify this model, at least from a statistical viewpoint, Gladman et al. (1997) numerically simulated the evolution of several hundred fictitious particles initially placed into the ν_6, 3/1, 5/2, 7/3, 2/1, 9/4 and 8/3 resonances. The computations have been done using Levison and Duncan (1994) SWIFT integrator, which is based on Wisdom and Holman (1991) symplectic algorithm but with substantial modifications in order to deal with planetary close encounters. Thanks to the efficiency of the integration code and to improvements in computer technology, the simulations covered a timespan of 100 Myr, namely two orders of magnitude longer than the previous simulations (Froeschlé et al. 1995). This allowed to follow the evolution of almost all the test particles from their source region up to their dynamical end-state (Sun-collision, planetary collision or ejection from the Solar System), thus deriving quantitative results concerning their lifetimes and the relative frequency of the possible end-states.

6.1 Dynamical Lifetimes

The basic result in Gladman et al. (1997) is that the median lifetime of the objects originally placed in the main resonances (ν_6, 3:1 or 5/2) is very short ($\sim$ 2 Myr) while only 10% of the particles manage to live more than $\sim$ 10 Myr (Fig. 11).

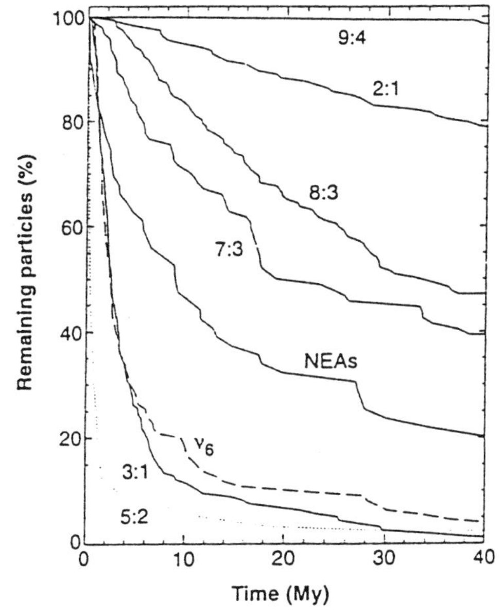

Fig. 11. Decay of the number of active particles vs. time. The ν_6, 3/1 and 5/2 resonances are characterized by very short lifetimes for most resonant particles. The outer resonances (7/3, 9/4 and 2/1) also have short times after the particle finds the "strongly chaotic" portion of the resonance, but finding these strong regions typically requires tens of millions of years. The decay curve of the known NEAs is shown for comparison. The population of known NEAs is biased towards particles on long-living orbits, so that its decay rate should not be taken to indicate any particular resonance as their source (from Gladman et al. 1997).

This is due to the fact that these resonances are very efficient in pumping the eccentricity to 1 on a Myr timescale, forcing the objects to collide with the Sun. Therefore, only the particles which are extracted from the resonances by the inner planets could live in principle significantly longer than 1Myr. By 'extraction' we mean a close encounter which moves the semi-major axis enough to take the particle out of the resonance. Mars is able to extract only a few percent of resonant bodies, which, after a typical time ranging from 1 to a few 10 Myr, always find themselves again re-injected into some resonant mechanism which pumps them to Earth-crossing orbits. The Earth and Venus are much more efficient than Mars in extracting bodies from resonances. However, the "mortality" of the extracted particles is still very high. Most of those which are driven by close encounters to $a > 2.5$ AU are ejected by Jupiter on hyperbolic orbits. Many others are injected again into the 5:2, 3:1 or ν_6 resonance and subsequently are forced to collide with the Sun. Only the minority of the objects which reach semimajor axes <2 AU live significantly longer. In fact, in this region, although many resonances exist and influence the dynamics (Michel and Froeschlé 1997), no statistically significant dynamical mechanisms have been found which pump the eccentricities up to Sun-grazing values. Therefore, particles with $a < 2$ AU can be eliminated only after being driven back to $a > 2$ AU (after a typical 1-10 Myr journey) and being pushed by a resonance into the Sun. Fig. 12 shows the typical evolutions of long-living particles, while Table 1 resumes several statistical aspects of the evolution of the particles, including the relative frequency of the different end-states.

The resonances beyond the 5/2 mean motion commensurability (7/3, 2/1, 9/4) are very inefficient sources of NEAs and meteorites since, at such large semi-major axis ($a > 2.8$ AU) as soon as the particles become Earth-crosser they have evolutions dominated by Jupiter's encounters and are quickly ejected from the Solar System.

Fig. 12. (*facing page*) Orbital evolutions of objects leaving the main belt. The curves indicate the set of orbits having aphelion (Q) or perihelion (q) at the semi-major axis of one of the planets Venus, Earth, Mars or Jupiter. **A** Common evolutionary paths from ν_6 and 3/1 resonances (eccentricity pumped to 1 and collision with the Sun) and 5/2 (ejection by Jupiter encounters. **B** one of the rare long-living particles from the ν_6; points are dots before 20 Myr and square after. **C** one of the rare long-living particles from the 3/1. Points are dots before 15 Myr and squares after (from Gladman et al. 1997).

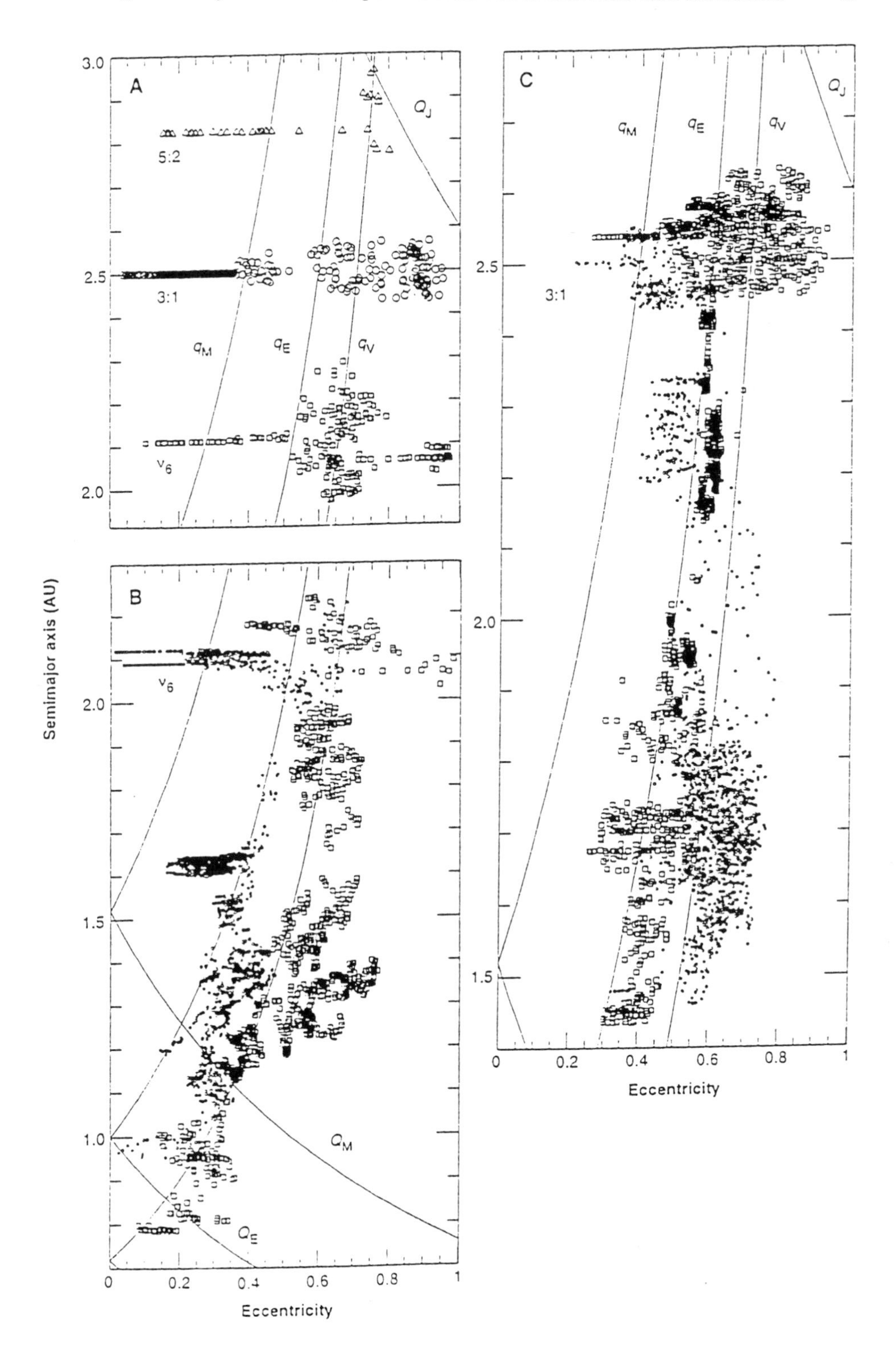
A
B
C
5:2
3:1
ν_6
q_M
q_E
q_V
Q_J
Q_M
Q_E
Semimajor axis (AU)
Eccentricity
3.0
2.5
2.0
1.5
1.0
0
0.2
0.4
0.6
0.8
1

6.2 Steady State and Asteroid Showers

At the light of the newly discovered short dynamical lifetimes, is it still reasonable to expect that the NEAs population is in a sort of steady state?

Farinella et al. (1993b) investigated which asteroids in the main belt could inject several large-sized fragments into the ν_6 or the 3/1 resonance. On the basis of their result, Menichella et al. (1996) estimated that ~ 400 asteroids, 1 km in size, should be injected per million year in such resonances, the number $N(R)$ of NEAs larger than a given size R being proportional to R^{-3}. These numbers seem to justify a positive answer to the above question, for what concerns Near Earth Asteroids smaller than 1 km. Conversely, the injection into resonances of bodies much larger than 1 km should be a too rare event to expect the existence of a steady state population of multi-kilometer Near Earth Asteroids. Unfortunately, at least two bodies among the presently known NEAs have a size larger than 10 km (433 Eros and 1036 Ganymed), this fact pointing to a still not complete understanding of the mechanisms at the origin of NEAs (see Sect. 7 for discussion).

Even if the NEAs population will be confirmed by future works to be basically in steady state, nevertheless the history of the Solar System should have been characterized by sporadic periods of NEAs overproduction. Zappalá et al. (1997) showed that several asteroid families should have injected a large number of family members into the neighbouring resonances at the moment of their formation (see also Morbidelli et al. 1995). The estimated number is $\sim 10^4$ for 1-km sized asteroids which, compared to the 400 bodies/Myr injected by the background asteroids (Menichella et al. 1996), indicates that a family break-up should be followed by a sharp increase of the number of NEAs as well as of the number of asteroidal impacts on the Earth. However, since only $\sim 10\%$ of the injected bodies survive ~ 10 Myr (Gladman et al. 1997), the number of NEAs should decrease back to its steady state value on a timescale of order ~ 10Myr; the latter should therefore be the typical timescale of such "asteroid showers" on our planet.

Table 1. (*facing page*) Summary of the numerical simulations. Resonant injections are labeled by the asteroid family, and the fraction of particles suffering each fate of surviving the entire integration is given. Time scale are shown for 50% (half-life) and 90% decay of the active particles, and median time T_{cr} for crossing the orbits of Mars, Earth and Venus. We list the percentage of particles that undergo extraction by Mars or by Earth and Venus combined (E/V), go directly into the Sun before an extraction event, or acquire certain orbits, with the median time spent in those orbits (T_{med}). Dashes indicate no simulated particles evolved in that way. The semimajor axis a_{res} of the center of the ν_6 resonance depends on the inclination: $a_{res} \simeq 2.1$ astronomical units (AU) for $i = 7^o$. The evolution statistics for the 8:3 resonance are based on only the first 29 million years (My) of the simulation. Q indicates orbital aphelion.

	Vesta		Nysa	Maria	Chloris	Dora	Gefion	Koronis		Eos	Themis
	ν_6	3:1	3:1	3:1	8:3	5:2	5:2	5:2	7:3	9:4	2:1
a_{res} (AU)	≃2.1	2.50	2.50	2.50	2.71	2.82	2.82	2.82	2.96	3.03	3.28
Mean initial i (degrees)	7	6	3	15	10	8	9	2	2	9	1
Length (My)	100	18	100	61	42	100	100	17	40	120	100
Active particles (n)	110	92	145	156	157	152	146	84	94	134	153
					End states (%)						
Impact sun	79.1	75.0	69.7	69.8	17.2	7.9	7.5	6.0	0	2.2	6.5
Impact Venus	1.8	0	0	0.6	0	0	0	0	0	0	0
Impact Earth	5.5	0	0.7	0	0	0	0	1.2	0	0	0
Impact Jupiter	0	0	0.7	0.6	0	0	0	4.8	1.1	0	0.7
Outside Saturn	11.8	25.0	28.3	28.8	35.7	92.1	92.5	88.1	56.6	25.4	22.9
Survivors	1.8	0	0.7	0	47.1	0	0	0	42.3	72.4	69.9
					Time scales (My)						
Half-life	2.3	2.1	2.6	2.5	34.0	0.4	0.6	0.7	19.0	>120	>100
90% decay	21.0	6.4	7.3	11.4	>42	3.4	2.9	3.8	>40	>120	>100
T_{cr} Mars	0.24	0.20	1.0	0.32	10.1	0.02	0.13	0.28	10.0	79.5	22.5
T_{cr} Earth	0.50	1.1	1.4	0.96	11.0	0.22	0.41	0.63	6.3	90.0	27.8
T_{cr} Venus	0.70	1.3	1.9	1.2	13.0	0.31	0.45	0.67	2.2	87.9	31.6
					Evolution (%)						
Extracted by Mars	8.2	3.3	5.5	4.5	16.6	–	0.7	–	42.6	30.5	3.3
Extracted by E/V	87.3	66.3	66.2	70.5	32.5	99.3	99.3	100	4.3	0.7	21.6
Direct to sun	4.5	30.4	28.3	25.0	–	0.7	–	–	–	–	3.3
Ever have $a < 2$ AU	53.6	8.7	9.0	7.7	0.6	0.7	1.4	3.6	–	0.7	–
T_{med} (My)	1.5	0.27	0.5	1.4	2.9	39.0	0.6	0.1	–	5.8	–
Ever have $a < 1$ AU	9.1	–	0.7	0.6	–	–	–	–	–	–	–
T_{med} (My)	2.8	–	0.04	13	–	–	–	–	–	–	–
Ever have $Q < 2$ AU	12.7	–	2.1	1.3	–	0.7	–	–	–	0.7	–
T_{med} (My)	6.3	–	0.2	13	–	2.0	–	–	–	4.1	–

6.3 Orbital Distribution

Assuming that the present epoch is not dominated by an "asteroid shower" but is basically in steady state situation, the numerical simulations by Gladman et al. (1997) allow to estimate the expected orbital distribution of NEAs. With respect to previous computations based on oversimplified models (see for instance Rabinowitz 1997) a large fraction of the simulated particles (one third of those which reach $a < 2$ AU) is found at large inclination ($I > 25°$). This is due to the 3/1 resonance, which also pumps the inclination, and to several secular resonances (including ν_{16}) which the NEAs encounter during their evolution. Therefore, the origin of the observed high-I NEAs should not be considered as mysterious. A few simulated particles are also found to be temporarily on Earth-like orbits, like those of the observed SEAs (Small Approaching Asteroids – Rabinowitz et al. 1993).

Unfortunately, a more quantitative comparison with the orbital distribution of the observed NEAs is not straightforward, because of the several discovery biases whose modelling is controversial.

Conversely, the comparison with the observed orbital distribution of fireballs of meteoritic origin (Wetherill and ReVelle 1981; Halliday et al. 1996) is much easier. In fact, meteorites can be considered as small NEAs (size ranging from few centimeters to about 1 meter), whose discovery is biased by the collision probability with the Earth and by an atmospheric entry velocity cutoff at about 20 km/sec (bodies entering the atmosphere at higher velocity should not deliver a meteorite at ground – Wetherill and ReVelle 1981). Morbidelli and Gladman (1997) have concluded that the orbital distribution of fireballs is in very good agreement with the one expected in a scenario of steady state injection into both 3/1 and ν_6 resonances and assuming an injection rate 5 times larger for the 3/1 resonance (Fig. 13).

7 Conclusions

The improved knowledge of the dynamical evolution of bodies emplaced in the main resonances of the asteroid belt has allowed a better understanding of the origin of both Near Earth Asteroids and meteorites.

However, two major problems still challenge our scenario:

1) the origin of multi-kilometer NEAs
2) the cosmic ray exposure age of meteorites.

As discussed in Sect. 6, the injection into resonances of multi-kilometer asteroids is a rare event, so that, being the lifetime of resonant particles very short, we would not expect multi-kilometer asteroids in the present NEAs population. This is evidently in contrast with the observations. It is possible that big NEAs do not come from the main resonances of the asteroid belt, but rather from the high-e portion of the asteroid belt, transiting through a Mars-crossing phase

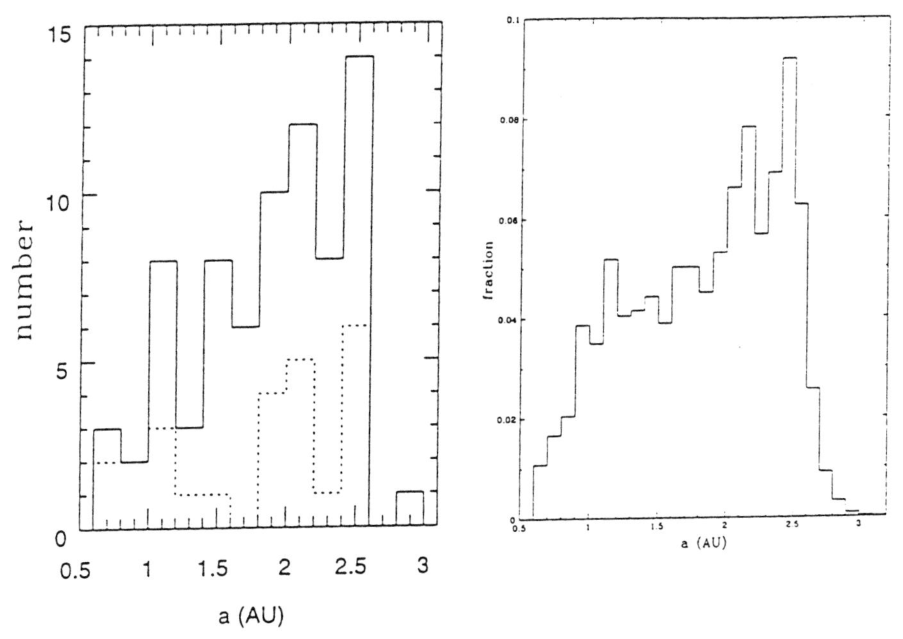

Fig. 13. Left panel: the binned semi-major axis distribution of the observed fireballs thought to be chondritic objects capable to drop meteorites. The solid histogram refers to data from both Wetherill and ReVelle (1981) and Halliday et al. (1996); the dashed histogram refers to Wetherill and ReVelle sub-sample. Right panel: the predicted orbital distribution of fireballs assuming a constant injection rate of bodies 5 times higher in the 3/1 than in the ν_6 resonances; a collisional lifetime of $\sim$ 3 Myr is also assumed (Morbidelli and Gladman 1997).

(Migliorini, private communication). How this could be possible (collisions, dynamical phenomena related to high order resonances, ...) is still not clear and deserves careful investigation.

The cosmic ray exposure age of meteorites measures the time interval in space between the meteoroid's formation as a body of <3 m diameter (following removal from a shielded location within a larger object) and its arrival at the Earth. The exposure ages of chondrites are typically of order 10-30 Myr, while those of iron meteorites are of order 10^8-10^9 y, in evident contrast with the typical lifetimes of particles after their injection into resonance (2 Myr). Morbidelli and Gladman (1997) have concluded that, in order to explain both the orbital distribution of meteorites and their exposure age, it is necessary to assume that meteorites are not directly injected into resonance by the event that removes them from shielded positions within their parent body, and that they are drifted into resonance by "some" mechanism on timescales comparable to their exposure age.

Farinella et al. (1997) have proposed that such mechanism could be provided by the so-called Yarkovsky effect (Opik 1951; Rubicam 1995). The Yarkovsky effect is produced by a non-conservative force, due to the non-isotropic thermal re-emission of the meteoroid. It causes slow changes in the meteoroid's semi-major axis.This could allow the capture into resonance of meteoroids, millions of years after their ejection from the parent body. However this scenario still requires further investigation.

Until a clear understanding of the origin of multi-kilometer NEAs and of the long exposure age of meteorites is not achieved, we cannot be sure that our scenario for the transport of asteroid material from the belt to the Earth is still not missing some fundamental aspect.

Dedication. This paper is devoted to our friend and collaborator Fabio Migliorini, who tragically died on Nov. 3 1997 during a hike in the Alps. He was 25. We will allways remember the enthusiasm he had in investigating the various facets of the NEAs origin puzzle.

References

Dermott, F.S., Murray, C.D. (1983): Nature of the Kirkwood gaps in the asteroid belt. Nature **301**, 201.

Farinella, P., Froeschlé, Ch., Gonczi, R. (1993a): Meteorites from the asteroid 6 Hebe. Celest. Mech. **56**, 287.

Farinella, P., Gonczi, R., Froeschlé, Ch., Froeschlé, C. (1993b): The injection of asteroid fragments into resonances. Icarus **101**, 174.

Farinella, P., Froeschlé, Ch., Froeschlé, C., Gonczi, R., Hahn, G., Morbidelli, A., Valsecchi, G.B. (1994): Asteroids falling onto the Sun. Nature **371**, 315–317.

Farinella, P., Vokrouhlický, D., Hartmann, W.K. (1997): Meteorite delivery via Yarkovsky orbital drift. Icarus [in press].

Froeschlé, Ch., Scholl, H. (1989): The three principal resonances ν_5, ν_6 and ν_{16} in the asteroidal belt. Celest. Mech. **46**, 231.

Froeschlé, Ch., Morbidelli, A. (1993): The secular resonances in the Solar System. In Milani A., Di Martino M., Cellino A. (eds.) *Asteroids, Comets, Meteors 1993*, Kluwer, pp. 189–204.

Froeschlé, Ch., Hahn, G., Gonczi, R., Morbidelli, A., Farinella, P. (1995): Secular resonances and the dynamics of Mars crosssing and Near-Earth asteroids. Icarus **117**, 45–61.

Gladman, B.J., Migliorini, F., Morbidelli, A., Zappalá, V., Michel, P., Cellino, A., Froeschlé, Ch., Levison, H. F., Bailey, M., Duncan, M. (1997): Dynamical lifetimes of objects injectd into the asteroid belt resonances. Science **277**, 197–201.

Halliday, I., Griffen, A.A., Blackwell, A.T. (1996): Detailed data for 259 fireballs from the Canadian camera network and inferences concerning the influx of large meteoroids. Meteoritics and Pl. Sci. **31**, 185–217.

Kirkwood, D. (1866): In *Proceedings of the American Association for the Advancement of Science for 1866.*

Kozai, Y. (1962): Secular perturbation of asteroids with high inclinations and eccentricities. Astron. J. **67**, 591–598.

Levison, H., Duncan M. (1994): The long-term behavior of short-period comets. Icarus **108**, 18–36.

Menichella, M., Paolicchi, P., Farinella, P. (1996): The main belt as a source of Near Earth-Asteroids. Earth, Moon and Planets **72**, 133–149.

Michel, P., Froeschlé, Ch., Farinella, P. (1996): Dynamical evolution of NEAs: close encounters, secular perturbations and resonances. Earth, Moon and Planets **72**, 151–164.

Michel, P., Froeschlé, Ch. (1997): The location of secular resonances for semimajor axes smaller than 2 AU. Icarus **128**, 230–240.

Moons, M., Morbidelli, A. (1994): Secular resonances in mean motion commensurabiltities: the 4/1, 3/1, 5/2 and 7/3 cases. Icarus **114**, 33–50.

Morbidelli, A., Moons, M. (1983): Secular resonances in mean motion commensurabilities: the 2/1 and 3/2 cases. Icarus **103**, 99.

Morbidelli, A., Zappalá, V., Moons, M., Cellino, A., Gonczi, R. (1995): Asteroid families close to mean motion resonances: dynamical effects and physical implications. Icarus **118**, 132–154.

Morbidelli, A., Gladman, B.J. (1997): Orbital and temporal distribution of meteorites originating in the asteroid belt. Meteoritics and Pl. Sci. [submitted].

Opik, E.J. (1951): Collision probabilities with the planets and the distribution of interplanetary matter. Proc. Roy. Irish Acad. **54**, 165.

Opik, E.J. (1976): *Interplanetary encounters*, Elsevier.

Rabinowitz, D., Gehrels, T., Scotti, J.V., McMillan, R.S., Perry, M.L., Wisniewski, W., Larson, S.M., Howell, E.S., Mueller, B.E.A. (1993): Evidence for a Near-Earth asteroid belt. Nature **363**, 704–706.

Rabinowitz, D. (1997): Are main belt asteroids a sufficient source for the Earth-approaching asteroids? Predicted vs observed orbital distribution. Icarus **127**, 33–45.

Rubincam, D.P. (1995): Asteroid orbit evolution due to thermal drag. J. Geophys. Res. **100**, 1585–1594.

Yoshikawa, M. (1989): A survey on the motion of asteroids in commensurabilities with Jupiter. Astron. Astrophys. **213**, 436.

Yoshikawa, M. (1990): On the 3:1 resonance with Jupiter. Icarus **87**, 78.

Yoshikawa, M. (1991): Motions of asteroids at the Kirkwood gaps. II. On the 5:2, 7:3 and 2:1 resonances with Jupiter. Icarus **92**, 94.

Wetherill, G. W., ReVelle D.O. (1981): Which fireballs are meteorites? A study of the Prairie network photographic meteor data. Icarus **48**, 308–328.

Williams, J. G. (1969): Secular perturbations in the Solar System. *Ph.D. dissertation*, University of California, Los Angeles.

Wisdom, J. (1983): Chaotic behavior and the origin of the 3/1 Kirkwood gap. Icarus **56**, 51.

Wisdom, J. (1985): Meteorites may follow a chaotic route to Earth. Nature **315**, 731.

Wisdom, J. (1986): *Chaotic dynamics in the Solar System.* Urey Lecture, November 1986, Paris, France.

Wisdom, J., Holman M. (1991): Symplectic maps for the N-body problem. Astron. J. **102**, 1528–1538.

Zappalá, V., Cellino, A., Gladman, B.J., Manley, S., Migliorini, F. (1997): Asteroid showers on Earth after family break-up events. Science [submitted].

Topics on Chaotic Transport for Hamiltonian Systems – Modelling of Diffusion Processes for Small Bodies in the Solar System

Claude Froeschlé and Elena Lega

O.C.A. Observatoire de Nice, B.P. 229, F-06304 Nice Cedex 4, France

Sur les transports chaotiques pour les systèmes hamiltoniens – Modélisation des processus de diffusion pour les petits corps du Système Solaire

Résumé. Des trajectoires chaotiques sont nécessaires pour amener au voisinage de la Terre des matériaux en provenance aussi bien de la Ceinture des Astéroïdes que des orbites cométaires. Pour étudier le comportement chaotique de ces petits corps, la modélisation est devenue à la fois un outil essentiel et un but en soi. Les équations de diffusion, les "mappings de Monte-Carlo" et les processus de Markov en sont les principaux instruments, que nous allons développer dans ce chapitre; nous examinerons également la puissance et les faiblesses de ces méthodes, en fonction de l'origine du chaos.

Abstract. Chaotic routes are essential to bring either asteroidal material or comets in the vicinity of Earth. To study the chaotic behaviour of asteroidal and cometary orbits, modelling becomes an aim and a tool. Diffusion equations, Monte-Carlo mappings and Markov processes are the main tools which will be discussed in this chapter as well as the strength and weakness of these methods versus the origin of chaos.

1 General Introduction

For centuries the aim of scientists working in celestial mechanics has been to solve Newton's equations which govern motions in the Solar System. More precisely under the assumption that all systems are integrable by a sequence of canonical transformations the part of the Hamiltonian we are unable to handle analytically is pushed to higher and higher order. This approach culminated with the work of Poincaré who showed that the solutions of the n-body problem could be formally written as series of purely periodic terms thus solving at least apparently the old controversy about whether secular variations of the semi-major axis occur at any order of perturbation theory. But he also showed that this series were generally asymptotic series and were divergent due to the appearance of uncontrolled small divisors.

Actually until the 1960's Poincaré discoveries were not fully appreciated. The KAM theorem on the analytical front and the semi-numerical studies of Hénon and Heiles (1964) show the coexistence of regular quasiperiodic motions and wild chaotic zones. These "Janus like" two faces of Hamiltonian systems are at the origin of a new understanding and new interest for dynamical problems related to the physic evolution of the Solar System. Indeed we are here at a third step of dynamical study of planetary science. The first one could be called the regular analytical approach. The second is connected with the new paradigm of dynamical systems and chaos. The third one being the interrelations between physics and dynamics especially for small bodies in the Solar System. Actually asteroid families are good examples of such an interrelation between : a physical hypothesis on the origin of a family by catastrophic encounter, the search for quasi first integrals under the assumption of regularity and the use of reliable cluster analysis methods (wavelet analysis, hierarchical clustering).

On the other hand, asteroidal material falling on the Earth and originating from such a catastrophic event is brought on Earth crossing orbits through a chaotic process. For carrying out such studies the usual tools of celestial mechanics break down and methods are used coming from other fields like mappings and modelisations of Markov process.

2 Stochastic Mappings in Astrodynamics

2.1 Introduction

Both the existence of Kirkwood gaps and the transfer of comets into observable orbits have been the main motivations for building mappings since they have many advantages over numerical integrations mainly with respect to computing time and to accuracy. However, even if the brutal force of super computer may in the near future solve these problems, building mappings will still be a challenge to researchers in Celestial Mechanics. Indeed to build mappings brings a deeper understanding of the general behaviour of non linear dynamical systems. In the particular case of mappings devised to study cometary motion , the origin of chaos is the key point.

2.2 Exogenous Stochastic Mappings

A Simple Model of Diffusion. It is well known that the study of dynamical systems with n degrees of freedom can be reduced to the study of a $2n-2$ measure preserving mapping, using the method of surface of section.

Here we deal with an area preserving mapping, but in which a stochastic element has been introduced (see Froeschlé 1975). Hence, we have researched how the classical situation of dynamical systems with two degrees of freedom was modified.

Such an approach, although academic, is interesting. It has been developed in a previous paper (Froeschlé, 1995) and we refer the reader to it.

Monte Carlo Mapping of Long-Period Comet Dynamics. While it has recently been realized that an important part of the dynamics of Oort cloud comets arises from regular motion in the Galactic tidal field (Heisler and Tremaine 1986), a decisive role is nonetheless played by individual stellar encounters. Given the physical parameters of such an encounter, its effect on the cometary motion is fully determined. However, the parameters of individual stellar encounters are unpredictable so the stellar perturbations impose a stochastic variation on the cometary orbits.

In Monte Carlo simulations of stellar perturbations (see, e.g., Weissman 1982; Remy and Mignard 1985) the dynamical evolution of a cloud of comets is studied as follows. At a given starting epoch, each comet is initialized by choosing a set of orbital elements. These are perturbed by the gravitational effect of passing stars. The geometrical parameters of the stellar encounters are chosen at random. During the passage of a star a comet receives a heliocentric impulse through the interaction of the star with the comet and the Sun. This induces a change in the cometary orbital elements, so these are updated and the comet moves along a new Keplerian ellipse until the next encounter with a random star. In Weissman's procedure the impulses are taken in an even more simplified way from a pre-determined distribution. In other words a stochastic mapping is iterated where the perturbations caused by random stars impose a stochastic process on the cometary orbital elements, which, therefore, undergo a random walk. It is obvious that all orbits are chaotic and correspondingly the largest LCE is strictly positive. The stochasticity is exogenous since the stellar encounters occur at random.

2.3 Endogenous Stochastic Mappings

Monte Carlo Mapping of Short-Period Comet Dynamics. As comets are captured into short-period orbits, i.e. with orbital period of the same order as those of the perturbing planets, the situation changes in a fundamental way. We now have to deal with only the intrinsic stochasticity of a dynamical system which can usually be approximated by a three-body problem: Sun-planet-comet (Rickman and Froeschlé 1988). This stochasticity derives mainly from close encounters with the planet, and during the intervals between such encounters the cometary motion is quasi-regular and predictable. In more precise terms, this means that the motion is stochastic only over time scales longer than the typical interval between encounters with the perturbing planet. Thus the phase space domain of short-period comets presents chaotic and ordered regions in an intricate mixture.

Stochastic modelling of this motion is justified by the shadowing principle only if the time step is chosen long enough. Thus, we may consider the following procedure (Rickman and Vaghi 1976; Froeschlé and Rickman 1980): we pick a large number of initial cometary orbits at random into boxes of the (Q, q) plane (Q aphelion and q perihelion distances as shown on Fig. 1 and integrate these with Jupiter as the perturbing planet over a time Δt, which was chosen to be one unperturbed period. From the resulting set of perturbations, we construct stochastic orbital evolutions by picking independently up, a random perturbation, within a corresponding box, for each successive interval Δt. For these evolutions, in order to simulate real ones, we would have a constraint on the choice of Δt so that a minimum time step for the random walk could in principle be defined. However, there are practical problems of such a procedure which are not easy to solve. These are connected with the question of the definition of the perturbation samples: how should one choose the "boxes" of phase space where the initial conditions are to be picked up ? As already mentioned, the dimension of the boxes can be reduced by symmetry arguments and in the present case a critical point is that short-period comet orbits are known to be generally of low inclination. One is therefore close to a planar problem with a four-dimensional phase space. Within the limits chosen for the inclination, the choice of inclination and orientation of the nodal line is statistically immaterial. Furthermore, since we are dealing both with Jupiter-crossing or Jupiter-tangent orbits and Jupiter's eccentricity is quite small, there is an approximately circular symmetry such that encounters with the planet can occur with equal probability, independently of the orientation of the apsidal line. The choice of the latter is hence also immaterial, and obviously the time-related parameter expressing Jupiter's position, at the time of the cometary perihelion passage, is the real stochastic variable of the problem which should be taken at random with its true probability distribution. We are left with two orbital parameters which can be taken as, e.g., semi-major axis and eccentricity $(a,\ e)$, or aphelion and perihelion distances $(Q,\ q)$.

As already noticed, boxes in the $(Q,\ q)$-plane were considered in the Monte Carlo simulations above mentioned. Let us note that a further reduction of the dimensions appears feasible. The Tisserand criterion for a co-planar cometary orbit :

$$T = \frac{2aJ}{Q+q} + 2\sqrt{\frac{2q\ q}{aJ(Q+q)}} = constant$$

might be used to restrict the random walk $(\Delta Q,\ \Delta q)$ to one-dimensional curves, and the Monte Carlo simulation would then consist of a set of independent, parallel simulations for the different values of T. In each of these one would consider a suitable orbital parameter varying along the curve, such as the inverse semi-major axis $z = 1/a$, and there would be a random walk with step size distribution $f_i(\Delta z)$ computed for interval $[z_{i-1},\ z_i]$ along the z axis. However, although small (Froeschlé and Rickman, 1981), the perturbations ΔT in the elliptic restricted problem are indeed important for the out-come of the low-velocity encounters with Jupiter occurring in low-eccentricity planet-tangent orbits, and those encounters appear essential for the dynamical transfer of comets.

Quite obviously the number of $(Q,\ q)$-boxes is limited in practice by the requirement to compute a sample of perturbations large enough to give a fair representation of the dynamics of the region in question. E.g., pronounced non-Gaussian tails are known to exist in the perturbation distributions and the sample must extent far enough into these tails to cover their significant parts. This means that the number of orbits to be integrated in each box may be very large

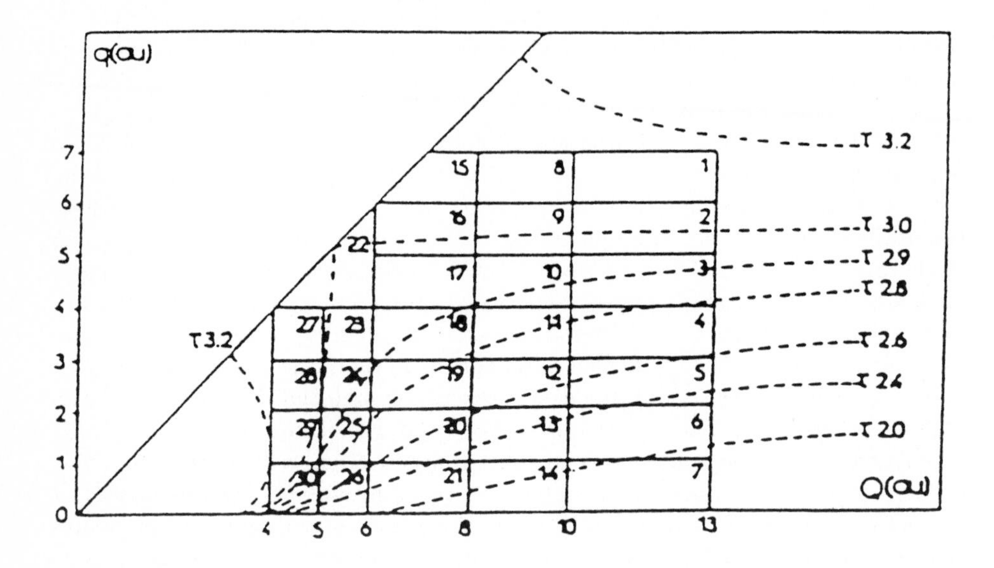

Fig. 1. The 30 regions of the $(Q,\ q)$-plane. Evolutionary curves for seven different values of the Tisserand constant are shown (dashed curves). From Rickman and Froeschlé (1979).

[for a possible way to reduce this problem, see Rickman and Froeschlé (1983)]. At this time it was difficult to go much beyond the dynamical resolution of the 30 boxes. On the other hand, this already guarantees a certain "dynamical homogeneity" in the sense that the perturbation distribution over a short interval of time (one orbital period) should not vary too much from one side of a box to the opposite one.

However, within all the boxes we can expect to find smaller regions corresponding to "resonant strips" with $a \approx a_{res}$, where the comets are close to a simple mean motion resonance with Jupiter. Within such a region there may be slow circulation of the critical argument, in which case it is extremely difficult to define a mean interval between encounters and to find an integration interval Δt which is everywhere sufficient. But actually the situation is even worse, because there also exist integrable regions of phase space corresponding to stable librations of the critical argument, where encounters with Jupiter never occur.

The main result of a Monte Carlo simulation of cometary dynamics is a picture of the distribution of comets over the various orbits connected by the stochastic transfer process in question. If the short-period comets are viewed in their most general framework, the dynamics includes perturbations by all the planets as well as non gravitational perturbations, and it would then be interesting to estimate the number of comets trapped in such quasi-stable resonant regions. How this goal would be achieved is not yet clear: the problem is to find an appropriate definition of the perturbation sample boxes and corresponding integration intervals for such a detailed investigation. If on the other hand, one considers the dynamical transfer in the three-body problem (Sun-Jupiter-comet), the problem is instead to reach what may be called "topological homogeneity" (Froeschlé and Rickman 1988): comets should then follow only chaotic routes and the relevant boxes from which the initial conditions are to be picked are in fact the intersections of the usual $(Q,\ q)$-boxes with the chaotic part of phase space. If this restriction is not taken into account, the rate of the transfer is artificially slowed down by inclusion of irrelevant trappings or an overestimated probability of small perturbations.

Markov Chain Modelling

Short-Period Comet Dynamics. If indeed we consider the orbital distribution of comets as the principal result to be obtained, we can again argue that the random walks by individual sample comets considered in the Monte Carlo simulation contain too much information: the only interesting quantity is the number of comets in each $(Q,\ q)$-box either as a function of time or in a steady state. The same perturbation samples used for the Monte Carlo simulation can then be used to calculate "jump probabilities" between the various $(Q,\ q)$-boxes over a common time interval Δt. Defining the state vector $\mathbf{n}$ of the cometary population to be the set of numbers of comets in the different boxes, and calling P_{ij} the jump probability from box i to box j, $\mathbf{n}$ evolves according to a Markov chain:

$$\mathbf{n}(t + \Delta t) = \mathbf{n}(t).P$$

where $P = (P_{ij})$ is the transformation matrix. This method was first used in cometary dynamics by Rickman and Froeschlé (1979) for the same domain of the $(Q,\ q)$-plane as in the above-mentioned Monte Carlo simulation and later on for Oort cloud dynamics by Lago and Casenave (1983).

This Markov method has the advantage of extreme efficiency, in particular for finding steady-state solution where we have just to solve :

$$\mathbf{n}_{ss} = \mathbf{n}_{ss}.P$$

i.e., a system of linear equations. Its main disadvantage is that, just like for the Monte Carlo simulation, the number of $(Q,\ q)$-boxes is limited by the large number of integrated orbits required to obtain accurate estimates of the jump probabilities. If e.g., the 30 boxes of Rickman and Froeschlé are considered, the information obtained on the $(Q,\ q)$ distribution is strictly limited to the 30 sample points represented in the state vector.

The Long-Term Dynamical Behaviour of Small Bodies in the Kuiper Belt. Again, like Rickman and Froeschlé (1979) but in different spirit, in order to study the slow diffusion of small bodies in the Kuiper belt, i.e. to determine the time scales of this process, Levinson (1991) has used Markov chain methods. Here we are not seeking any more steady states solutions but estimates of diffusion times which should be of the order of the age of the Solar System.

The $(Q,\ q)$ plane is also divided into small bins for which transition probabilities are estimated through sample of 100 particles in each bin which are integrated for approximately 100 periods. In addition to the bins within the Kuiper belt, which is a transient region, two special absorbing bins act as border edges, an inner edge $q < 30$ AU where the objects become Neptune crossers and an outer edge left as a free parameter with corresponding probability $P_{ii} = 1$ and $P_{ij} = 0$ for $i \neq j$.

The probabilities matrix takes the form:

$$P = \begin{pmatrix} I & O \\ R & Q \end{pmatrix}$$

and the fundamental matrix

$$M = (M_{ij}) = (I - Q)^{-1}$$

gives the average number of time steps a particle spends in transient bin j before it is absorbed if it started in bin i. Then $t_i = \sum_j M_{ij}$ gives the average time a particle spends in all transient bins before absorption. Furthermore, the variance of the particle lifetime is $v = (2M - I)t - s$ where $s_i = t_i^2$. Finally $f = MR$ gives the probability that a particle starting in transient bin i enters absorbing bin j. Using these tools Levinson found that the Kuiper belt is a good candidate as a source of short-period comets. However, in addition to the already discussed possible draw backs of the method (size number of bins, time steps length) the stochasticity underlying the Markov process is indeed very small and may only be due to undetected long-period oscillations in the behaviour of q and Q.

In the previous study such a low stochasticity was indeed present as an artifact of the method but in the study concerning short-period comets dynamics a strong stochasticity due to strong interactions between the comet and Jupiter was mainly responsible for the diffusion, i.e. big-jumps occurred within Δt and therefore such an artifact could not invalidate the results.

3 Another Approach: The Synthetic Mapping

The validity of stochastic mappings is not yet unquestionable since it depends obviously of the intrinsic stochasticity of the region of the phase space under study. An other method also, purely numerical, has been used by Froeschlé and Petit (1990, paper I) who built a mapping valid everywhere in the phase space, following an idea already used by Varosi et al. (1987) but in the framework of non-Hamiltonian systems (i.e., systems where attractors do exist). The method consists of coarse-graining the phase space surface of section and then interpolating the value of the image of a point. Linear interpolation requires a rather fine graining of the phase space, hence it is necessary to compute a lot of points on the grid. However, Taylor expansions of order 3 and 5 can provide very good results as long as symmetrical interpolation formulae are applied, for which it is necessary to use an extended grid. Since there are cases where one cannot cross a given limit, asymmetrical interpolation formulae have been tested, but their accuracy was found to be inferior. Therefore Petit and Froeschlé (1994, paper II) have developed another type of interpolation, where the information, including that on the gradients, is stored to the same level of accuracy only for the nearest-neighbouring vertices. Thus, not only images of vertices are computed, but also tangential mappings at each vertex.

In any case, the synthetic mapping obtained is not symplectic even if at vertices there is an exact interpolation; therefore we have introduced a second method (Froeschlé and Lega 1996) based on a global fitting. The polynomial are obtained using at once all the vertices and fitting by least square polynomes but in such a way that the symplectic character is not lost.

3.1 A Local Exact Fitting

Poincaré maps are now of common use for studying the qualitative behaviour of differential equations (see Hénon, 1981). Moreover, in order to study stability problems, many authors have sought explicit algebraic mappings which approximate, at least qualitatively, the Poincaré maps obtained from the original Newton equations. Froeschlé and Petit (1990) have reviewed some of these mappings and showed that they are reliable only as long as one remains within the domain of validity of the approximations made in order to isolate either - in the case of deterministic mappings - an integrable part and some instantaneous perturbations, or - for stochastic mappings - a source of endogenous/exogenous stochasticity (see Froeschlé and Rickman 1988). All these mapping are *ad hoc* and reliable only in some region of the phase space and for some specific purpose. In Froeschlé and Petit (1990) we built a mapping valid everywhere in the phase space, following an idea already used by Varosi et al. (1987) but in the framework of non-Hamiltonian systems (i.e. systems where attractors do exist). The method consists of coarse-graining the phase-space surface of section and then interpolating the value of the image points. Linear interpolation requires a rather fine graining of the phase space, hence it is necessary to compute a lot of points on the grid. However, Taylor expansion of order 3 and 5 can provide very good results as long as symmetrical interpolation formulae have been tested, but their accuracy was found to be inferior. Therefore Petit and Froeschlé (1994) have developed another type of interpolation, where the information, including that on the gradients, is stored to the same level of accuracy only for the nearest-neighbouring vertices. Thus, not only images of vertices are computed, but also tangential mapping at each vertex.

There are in any case two key parameters: the number of bins in each direction N =(total number of cells)$^{1/D}$, where D is the dimension of the surface of section, and the order M of the Taylor expansion. In order to explore the validity for the synthetic approach we have applied our method for an algebraic area-preserving mapping for which the computation of orbits is very fast. This allows one to follow a large number of orbits and to carry out enough iterations for a meaningful comparison. In this case we have used the well known standard mapping (Froeschlé 1970; Lichtenberg and Lieberman 1983) :

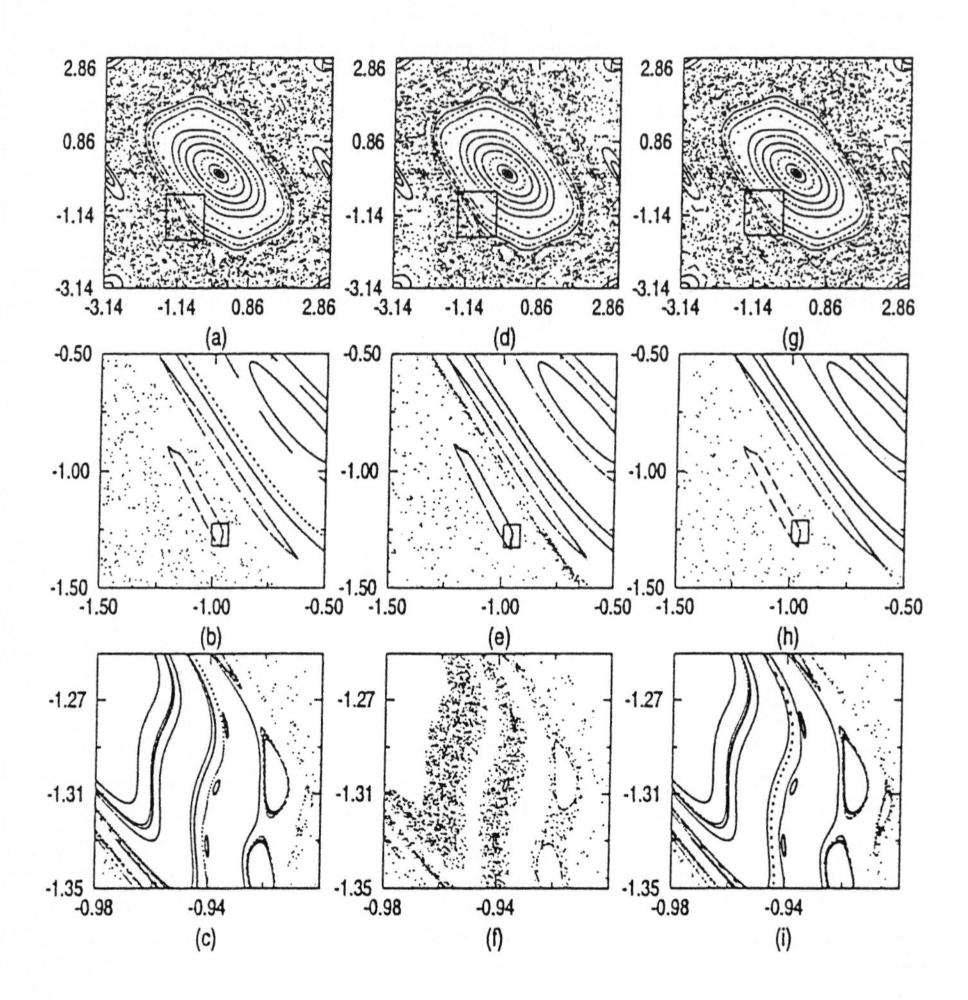

Fig. 2. (a) Plot of the standard map for a=-1.3, (b) and (c) are enlargements of the small boxes shown respectively in (a) and (b); (d), (e) and (f) are the same as (a-b-c) but using a Taylor approximation of order 5. (g), (h) and (i) : same as (a-b-c) but using the global method explained in Sect. 3.2.

$$\begin{cases} x_{i+1} = x_i + a\sin(x_i + y_i) \\ y_{i+1} = y_i + x_i \end{cases} \pmod{2\pi} \tag{1}$$

Figure 2a shows orbits of the standard mapping for $a = -1.3$. Indeed such a mapping exhibits all the well-known typical features of problems with two degrees of freedom, such as invariant curves, "islands", and stochastic zones where the points wander in a chaotic way. Figures 2b and 2c are magnifications of the small boxes indicated in Fig. 1a. At this magnification level, details like second-order islands become evident and the approximation levels of the synthetic maps are easily visualized. Figures 2d-e-f correspond to the same orbits and the same magnifications as Figs. 2a-c but using the Taylor interpolation mapping of order $M = 5$ (T5) and a number of bins $N = 20$. In order to get a finer description of the non-area preserving behaviour of T5 for an invariant curve, we have computed at regular time intervals (see Laskar et al. 1992) the rotation number using a linear interpolation method (M. Hénon, private communication). We have displayed (Froeschlé and Petit 1993) a contracting or expanding behaviour of T5 depending on the total number of cells and the starting point. This phenomena are a consequence of the non area preserving character of the local mapping. We present in the next section a new method based on a global area preserving mapping.

3.2 A Global Fitting: The Least Square Interpolation

In this new approach we look for the mapping $T^\star$

$$T^\star = \begin{cases} x_{i+1} &= P_m(x_i, y_i) \\ y_{i+1} &= Q_n(y_i, x_i) \end{cases} \tag{2}$$

where P_m and Q_n are polynomial of order m and n which fit the initial Poincaré map T (here the standard mapping) at the best (in the least square meaning) for the images of a regular grid. Let us emphasize that with this approach we loose the exact fitting at the vertices. It is the price we pay in order to obtain a symplectic mapping. In the two-dimensional case of $T^\star$ this means an area preserving mapping i.e. the determinant of the jacobian matrix is one. The problem of getting such polynomial expression is not a trivial one and we used as a basic synthetic mapping the following one:

$$\begin{cases} x_{i+1} &= x_i \\ y_{i+1} &= y_i + g(x_i) \end{cases} \tag{3}$$

where g is a polynomial of degree n obtained from a generating function by $g(x) = -\partial S/\partial x$. This is a shear along the y axis. Such a mapping is obviously symplectic as well as the following one:

$$\begin{cases} x_{i+1} &= x_i + h(y_i) \\ y_{i+1} &= y_i \end{cases} \tag{4}$$

where h is a polynomial of degree m. Since the problem is also to obtain a general form for the polynomial P_m and Q_n we have considered a composition of the mappings 3 and 4 which is still symplectic:

$$A = \begin{cases} x_{i+1} &= x_i + h(g(x_i) + y_i) \\ y_{i+1} &= y_i + g(x_i) \end{cases} \tag{5}$$

with

$$A^{-1} = \begin{cases} x_i &= x_{i+1} - h(y_{i+1}) \\ y_i &= y_{i+1} - g(x_{i+1} - h(y_{i+1})) \end{cases} \tag{6}$$

In the same way we can consider:

$$B = \begin{cases} x_{i+1} &= x_i + \alpha(y_i) \\ y_{i+1} &= y_i + \beta(x_i + \alpha(y_i)) \end{cases} \tag{7}$$

with

$$B^{-1} = \begin{cases} x_i &= x_{i+1} - \alpha(y_{i+1} - \beta(x_{i+1})) \\ y_i &= y_{i+1} - \beta(x_{i+1}) \end{cases} \tag{8}$$

We define $T^\star = A \circ B$ and considering that $T^\star(x_i, y_i) = (x_{i+1}, y_{i+1})$ we write $B(x_i, y_i) = A^{-1}(x_{i+1}, y_{i+1})$ in order to obtain the coefficients of the polynomes g, h, α, β using the least square fitting. Using this trick the computation is straight and does not require any algebraic manipulation. Of course we can also use $T^\star = B \circ A$ which will give different values for the coefficients. Figure 3 shows the variation of χ^2 (which measures the goodness of the fit) as a function of the polynomial degree (m and n) and of the number of fitted points ($N \times N$). We notice that the precision does not improve for $N \geq 20$ and we get minimum values respectively for $m = 13$ for the couple h, α and $n = 1$ for the couple g, β. The results on the mapping are shown in Fig. 2 (third column). Let us remark the very good agreement of Fig. 2c with Fig. 2i. In order to obtain the same result with the local fitting, it is necessary to increase of a factor 4 the total number of points $N \times N$. We have therefore to pay a higher price on computation time in comparison with the global method. Despite its global character, the global method ($T^\star$) ensures a better fit because of its area preserving properties.

3.3 Preliminary Conclusions

Synthetic maps appear to be valuable tools for celestial mechanics. We have presented here only some partial results. For instance another important development concerns problems with more than two degrees of freedom, for which the situation is less straightforward than described above. As far as the first method is concerned the number of operations required for the Taylor approximation increases drastically with the dimension of the surface of section. Of course a lower-order map can be used by decreasing the grid size, but a further difficulty lies in the task of storing and recalling the values of the computed images at the vertices. This is the reason why we have used a hash function when dealing with problems with three degrees of freedom (Petit and Froeschlé 1994). The restauration of simplecticity for P and Q appears to be promising since for the same accuracy we obtain at once a general fitting.

4 Conclusion

If the beautiful mathematical machinery of celestial mechanics is at the basis of the asteroid family story, the chaotic routes have however close connections with the finding of proper elements. Actually, the same machinery, but used in a complementary manner, gives the location and even the size of both mean motion and secular resonances and therefore the regions of the phase space where they overlap and generate chaos. Indeed slow chaotic diffusion, inducing increases of

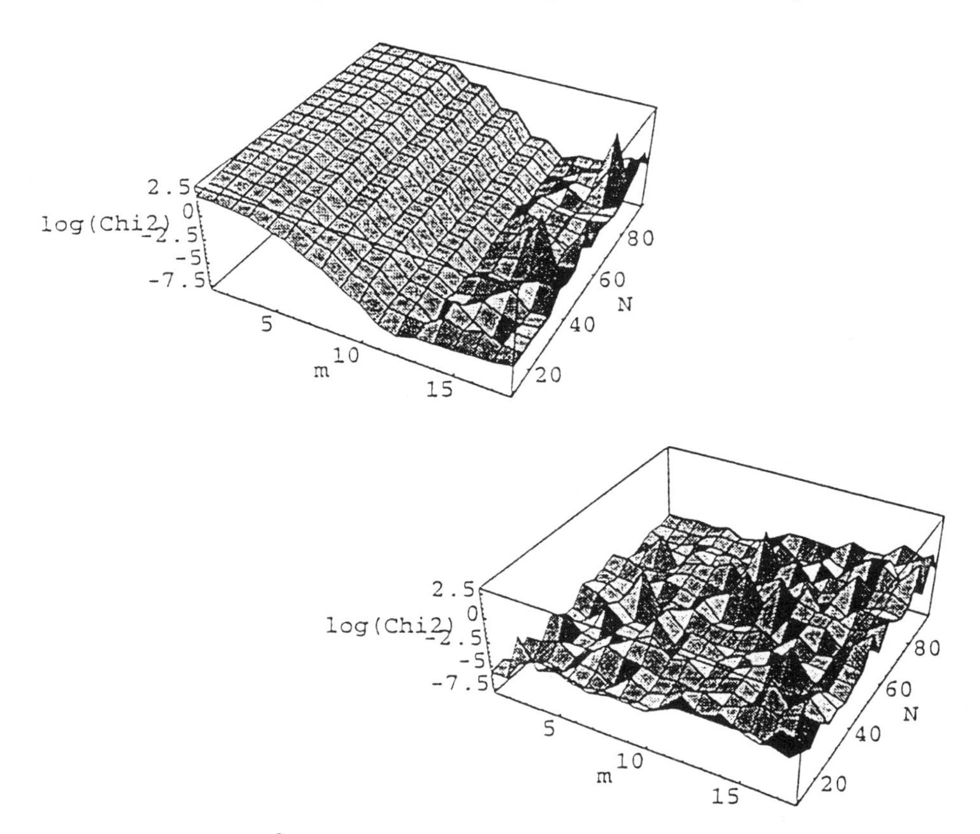

Fig. 3. Variation of χ^2 as a function of the polynomial degree m and of the number of fitted points $N \times N$. Upper left figure corresponds to the polynomial P, lower right figure corresponds to the polynomial Q.

eccentricities result of such resonant interactions and are good candidates for bringing material to the inner Solar System. This material must be understood as meteorites resulting of asteroid break up or as Kuiper belt comets. Close approaches with planets are at the origin of another source of chaos for which only crude models have been worked out so far. A better understanding of interactions between resonances and strong gravitational perturbations appears to be crucial for future models of chaotic routes through the Solar System.

References

Froeschlé, C. (1970): A numerical study of the stochasticity of dynamical systems with two degrees of freedom. A&A **9**, 15–23.

Froeschlé, C. (1975): Numerical study of a random dynamical system with two degrees of freedom. Astrophys. Space Sci. **37**, 87–100.

Froeschlé, C. (1995): Chaos in astrodynamics : modelling of diffusion processes for small bodies in the Solar System. In Benest D., Froeschlé C. (eds.) *Chaos and diffusion in hamiltonian systems*, Editions Frontières, pp. 199–221.

Froeschlé, C., Lega, E. (1996): Polynomial approximation of Poincaré maps for hamiltonian systems. Earth, Moon and Planets **75**, 51–56.
Froeschlé, C., Petit, J.M. (1990): Polynomial approximation of Poincaré maps hamiltonian systems. A&A **238**, 413–423.
Froeschlé, C., Petit, J.M. (1993) *Primo convegno nazionale di Meccanica Celeste*, L'Aquila (Italia).
Froeschlé, C., Rickman, H. (1980): New Monte Carlo simulations of the orbital evolution of short-period comets and comparison with observations. A&A **82**, 183–194.
Froeschlé, C., Rickman, H. (1981): A Monte Carlo investigation of jovian perturbations on short-period comet orbits. Icarus **46**, 400–414.
Froeschlé, C., Rickman, H. (1988): Monte Carlo modelling of cometary dynamics. *Celest. Mech.* **43**, 265–284.
Heisler, J., Tremaine, S. (1986): The influence of the galactic tidal field on the Oort comet cloud. Icarus **65**, 13–26.
Hénon, M. (1981): Numerical exploration of hamiltonian systems. In Iooss G, Helleman R.H.G., Stora R. (eds.) *Chaotic behaviour of deterministic systems – Les Houches, session XXXVI*, North Holland, pp. 54–170.
Hénon, M., Heiles, C. (1964): The applicability of the third integral of motion: some numerical experiments. A.J. **69**, 73–79.
Lago, B., Casenave, A. (1983): Dynamical evolution of cometary orbits in the Oort cloud : Another statistical approach. Icarus **53**, 68–83.
Laskar, J., Froeschlé, C., Celletti, A. (1992): The measure of chaos by the numerical analysis of the fundamental frequencies. Application to the standard mapping. Physica D **56**, 253–269.
Levinson, H.F. (1991): The long-term dynamical behavior of small bodies in the Kuiper belt. A.J. **102**, 787–794.
Lichtenberg, A.J., Lieberman, M.A. (1983): *Regular and stochastic motion*, Springer.
Petit, J.M., Froeschlé, C. (1994): Polynomial approximation of Poincaré maps hamiltonian systems II. A&A **282**, 291–303.
Remy, F., Mignard, F. (1985): Dynamical evolution of the Oort cloud I. A Monte Carlo simulation. Icarus **63**, 1–19.
Rickman, H., Froeschlé, C. (1979): Orbital evolution of short-period comets treated as a Markov process. A.J. **84**, 1910–1917.
Rickman, H., Froeschlé, C. (1983): A keplerian method to estimate perturbations in the restricted three-body problem. Moon and Planets **28**, 69–86.
Rickman, H., Froeschlé, C. (1988): Cometary dynamics. Celest. Mech. **43**, 243–263.
Rickman, H., Vaghi, S. (1976): A Monte Carlo simulation of the orbital evolution of comets in the inner planetary region. A&A **51**, 327–342.
Varosi, F., Gebogi, V., Yorke, J.A. (1987): Simplicial approximation of Poincaré maps of differential equations. Phys. Lett. A **124**, 59–64.
Weissman, P.R. (1982): Dynamical history of the Oort Cloud. In Wilkening L.L. (ed.) *Comets*, Univ. Arizona Press, pp. 637–658.

Collisional Disruption of Natural Satellites

Paolo Farinella

Dipartimento di Matematica, Università di Pisa
Via Buonarroti 2, I-56127 Pisa, Italia

Brisement collisionnel de satellites naturels

Résumé. Les observations des satellites naturels des planètes indiquent que nombre d'entre eux ont été probablement brisés par des impacts très énergétiques tout au long de l'histoire du Système Solaire. D'une part, la surface de plusieurs satellites, tant grands que petits, exhibe des cratères géants, indiquant des énergies d'impact proches du seuil au-delà duquel la cible se brise. D'autre part, beaucoup de petites lunes ont des formes irrégulières, qui suggèrent qu'il s'agit de fragments issus du brisement de corps primordiaux plus importants.
Les petits fragments éjectés lors d'un impact peuvent, dans certains cas, rester en orbite autour du corps "parent" et former des anneaux. Ensuite, ces fragments soit ont une longue durée de vie grâce à des mécanismes dynamiques de protection, soit se ré-acrètent pour former un nouveau satellite. Dans le cas d'Hypérion, la chute de nombreux fragments importants – issus de son brisement – sur Titan pourraient avoir été à l'origine de (ou du moins avoir substantiellement changé) la composition de l'atmosphère dense de ce grand satellite.

Abstract. The evidence indicating that many natural satellites have been disrupted by energetic impacts during the history of the solar system is briefly reviewed. Several satellites over a wide size range show giant craters close to the threshold for catastrophic breakup. Many small moons have very irregular shapes, suggesting that they are fragments from larger precursor bodies. Ejecta from the fragmentation of satellites in many cases may stay in planetocentric orbits, forming ring systems, reaccumulating into newly-born satellites or surviving thanks to dynamical protection mechanisms. In one satellite breakup case, that of Hyperion, the fall of a massive fragment shower onto Titan may have generated or substantially changed the composition of its dense atmosphere.

1 Introduction

Many natural satellites of the Solar System, in particular the smaller ones with strongly irregular shapes, are probably fragments formed after catastrophic impact events between larger precursor satellites and passing projectile bodies. This conclusion is suggested by several lines of evidence, which will be briefly reviewed below. For a more detailed discussion of specific issues, the reader is referred to the extensive literature appeared in recent years both in specialized planetary science journals and in some review books. Suitable references will be provided throughout the remainder of this chapter.

2 Giant Craters

On the surface of several satellites ranging in size between ≈ 1 and ≈ 1000 km we see giant craters, of diameter close to or exceeding the mean radius of the body, which were clearly produced by collisions of energy not much smaller than the threshold for catastrophic breakup (Chapman and McKinnon 1986; Fujiwara et al. 1993; Holsapple 1994). Two impressive examples very far away from each other are the Martian satellite Phobos (Fig. 1), whose Stickney crater is associated with an extensive pattern of "grooves" and has been the subject of extensive modelling work (Fujiwara 1991; Asphaug and Melosh 1993), and the Neptunian moon Proteus (Fig. 1), which has a mean diameter of about 400 km and an impact crater (Pharos) 250 km across and 10 to 15 km deep (Croft 1992).

Large craters are also evident in the Jovian, Saturnian and Uranian satellite systems (Schenk 1989; Croft 1992). Some examples are the following (in brackets I give the approximate ratio between the crater rim diameter and the satellite radius): Valhalla [0.58] on Callisto, Herschel [0.73] on Mimas, Odysseus [0.84] on Tethys, Arden [1.44] on Miranda. Although their morphology is very diverse, ranging from bowl-shaped craters on the smallest bodies, through flat-floored and concentric floored craters up to giant multiring basins for the largest moons, all these impact scars probably record the excavation and ejection of least a few percent of the target mass. It is also worth noting that many small, irregularly shaped satellites show large concavities or indentations which are not recognizable as well-defined craters, but have probably also been formed by quasi-catastrophic impacts (Thomas 1989).

From this type of evidence and from quantitative crater counts, already in the early 80's the Voyager imaging team (Smith et al. 1981, 1982, 1986, 1989) had realized that most of the moons of the outer planets – excluding the very largest bodies, namely the Galilean satellites, Titan and Triton – are the products of one or more catastrophic breakups of larger satellites some time in the past, with most bodies disrupted outside the Roche zone rapidly reaccumulating into newly-born moons with masses comparable to the initial satellites. In the outer solar system, an adequate population of potential impactors is provided by passing comets, of both short and long period (Shoemaker and Wolfe 1982; Plescia and Boyce 1983; Lissauer et al. 1988). Their flux and relative speed are significantly enhanced by the gravitational focusing effect caused by the presence of a nearby massive planet.

3 Irregular Shapes

All the natural satellites smaller than about 300 km have rugged limb topographies and irregular overall shapes, which can only roughly be approximated by triaxial ellipsoids (Farinella 1987; Thomas 1989). In this respect, the situation is similar to that of the asteroids, for which lightcurve data indicate that as a rule shapes are quite elongated and irregular (Catullo et al. 1984; Binzel et al. 1989; possible exceptions are the few largest asteroids). Close-up images obtained recently by the Galileo probe have also shown that there is a remarkable similarity in overall shape between the small, irregular satellites and the two small asteroids (951) Gaspra and (243) Ida, which are believed to be fragments from larger parent bodies disrupted by impact events (Ida is in fact a member of the Koronis family, for which an impact breakup origin is directly indicated by the similarity in the proper orbital elements).

Following the pioneering work of Fujiwara et al. (1978), the shape distribution of fragments from energetic impact or explosive events has been determined by several groups of experimenters, using both icy and rocky targets (Capaccioni et al. 1984, 1986; Lange and Ahrens 1987; Giblin et al. 1994). The most important finding from this work is that, when the fragment shapes are characterized by the ratios b/a, c/a between three axial dimensions $a > b > c$ along three perpendicular directions, these parameters are always distributed in a quasi-Gaussian way, with peaks at $b/a \approx 0.7$, $c/a \approx 0.5$ and standard deviations of about 0.15. This implies that both equidimensional fragments and highly elongated or flattened shapes are very rare. This property is quite insensitive to target composition and fragment size. Catullo et al. (1984) have used these results to show that main-belt asteroids smaller than about 100 km have typically fragment-like shapes. As argued by Farinella (1987) and Thomas (1989), the same can be said for the small satellites, although the number of them for which all the three axes are reliably determined does not exceed 10, and one must be cautious in generalizing from such a small sample. Actually, the smallest known satellite, asteroid Ida's Dactyl (diameter 1.4 km) is much less irregular in shape than a typical fragment.

In particular, among the satellites we probably see also some bodies (all larger than 100 km in diameter) which represent a transitional stage between the irregular, fragment-like shapes and the nearly-equilibrium shapes, moulded by self-gravitational, rotational and tidal forces. This is the case for Proteus, whose global figure appears to have relaxed to a nearly-equilibrium one, with $a \approx b \approx c$, but which also shows large-scale unrelaxed surface features. Similar to Proteus is probably Phoebe, the outermost Saturnian moon imaged (at low resolution) by Voyager 2. The overall shapes of the two coorbital Saturnian satellites Janus and Epimetheus are quite elongated, but much less irregular than those of the nearby F-ring shepherd satellites Prometheus and Pandora (Stooke 1992; Stooke and Lumsdon 1992) and not far from the equilibrium ones (assuming a density of about 1 g/cm^3), once the tidal distortion is taken into account (Farinella et al. 1985). The shapes of these bodies could be relaxed either due to internal heating and creep deformation (Johnson and McGetchin 1973), or because they are weakly consolidated agglomerations of former fragments rather than rigid, unfractured objects. Such "rubble pile" satellites could be the natural outcome of collisional fragmentation followed by reaccretion outside the Roche limit (Farinella 1987). This may be the explanation for Dactyl's nearly spherical shape (Davis et al. 1996).

A peculiar case is that of Hyperion (Fig. 1), which orbits Saturn between Titan and Iapetus. It is the only irregular satellite orbiting within groupings of much larger, nearly-spherical ones. It was found by Voyager 2 to spin about its *longest* axis with a large-amplitude wobble, and this suggested the possibility of a recent impact. However, later on Hyperion's rotational state was found to be chaotic by Wisdom et al. (1984), so irregular variations of both its spin rate and its polar axis are to be expected. With a mean radius of 135 km and axial ratios $c/a = 0.65$, $b/a = 0.79$, it is the largest, irregularly shaped icy object in the solar system. These properties suggest that the current Hyperion has been generated by impact fragmentation of a precursor object between 2 and 5 times the present volume (Farinella et al. 1983; Thomas et al. 1995). As we shall see later on, most of the fragments from this catastrophic breakup may have escaped reaccumulation due to the peculiarities of the orbital resonance locking between Hyperion and Titan (Colombo et al. 1974; Greenberg 1977).

4 Association with Rings

All the four ring systems of the giant planets are associated to a number of small, irregularly shaped satellites which orbit inside the rings themselves, near their edges or within gaps. The satellite-ring relationships and interactions are governed by a variety of subtle dynamical mechanisms, both gravitational (Franklin et al. 1984; Dermott 1984) and non-gravitational (Mignard 1984; Anselmo and Farinella 1984), which are not yet completely understood. However, it is clear that a plausible formation mechanism for these small satellites is collisional fragmentation of precursor moons inside or close to the Roche limit.

After the Voyager exploration of the outer planets, it has also been realized that a collisional fragmentation process may explain the formation of the rings themselves better than most alternative hypotheses (Harris 1984; Esposito 1993; Colwell 1994). The main reason is that the four planetary ring systems are different in nature, but they share the property that the typical time for their observable features to change significantly is much shorter than the age of the solar system. Therefore, they must be highly evolved systems. At the same time, the existence within or near the rings of a number of moons much smaller than many craters on the larger satellites (i.e., some 10 to 100 km in diameter) suggests that these objects cannot have survived 4.5 Byr of bombardment by comets and meteoroids, and that impact disruption of their parent bodies must have been the source of debris, which could take the form of visible rings. This hypothesis explains satisfactorily both the continued existence of rings with apparently short evolutionary lifetimes and the coexistence of moons and rings at comparable locations within the Roche limits of the giant planets.

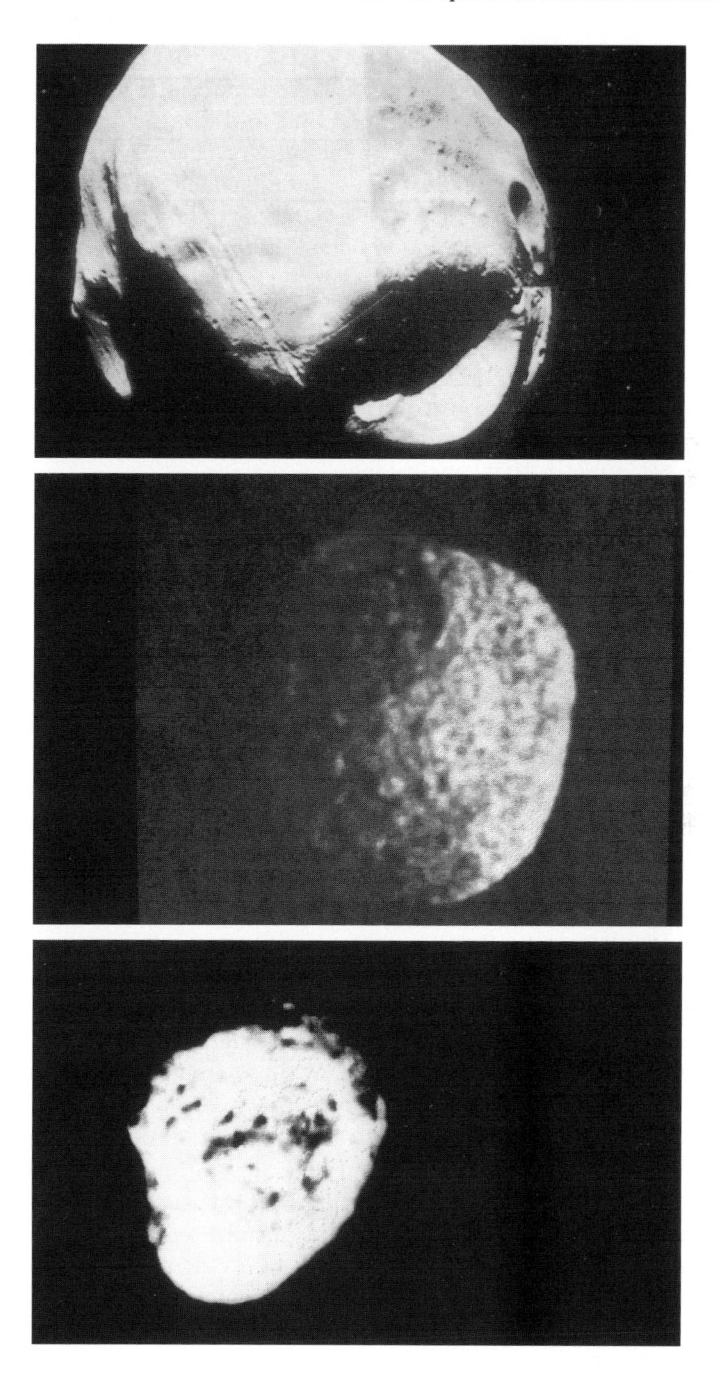

Fig. 1. A Viking image of the martian moon Phobos with crater Stickney (up), together with Voyager images of the neptunian moon Proteus with crater Pharos (middle) and of the irregular shape of the saturnian moon Hyperion (below) – Documents NASA.

Colwell and Esposito (1992, 1993) have recently developed detailed numerical simulations of the process creating planetary rings from disrupted satellites, in the specific case of the narrow rings of Uranus and Neptune. According to their results, circumplanetary bodies can undergo a "collisional cascade", leading to the creation of narrow rings and of moon populations similar to the currently observed ones. The model predicts that some observed moons are gravitationally bound "rubble piles" that have undergone multiple disruptions. Also, it requires the existence of many unseen km-sized moonlets, whose existence would provide adequate sources for the short-lived dust particles abundant in many rings. The absolute timescales of this process and the original moon populations remain uncertain, because the impactor flux and the impact strengths of the target satellites are poorly known. Moreover, the disruption-and-reaccretion cycle and the resonant interactions are so complex that much further modelling work is required to get quantitative conclusions.

5 Dynamical "Natural Selection"

Several small satellites are protected from destabilizing close encounters with larger bodies by resonant dynamical mechanisms, which suggest a kind of natural selection process preserving a few fragments out of a large initial population, thanks to their "lucky" dynamical configuration. An obvious example are the small Trojan moons of Dione, Tethys and possibly Mimas in the Saturnian system: like in the case of the Trojan asteroids, stable librations about the triangular Lagrangian equilibrium points prevent close encounters with the main satellites. The location of these moons clearly suggests that they are fragments from the disruption of precursor satellites, which did not reaccrete and could survive as individual objects owing to their peculiar dynamics.

Somewhat different is the case of the coorbital satellites Janus and Epimetheus, as they have comparable masses and follow horseshoe-type orbits, exchanging periodically their paths with respect to a common mean orbit. At each encounter between the satellites, energy and angular momentum are exchanged in such a way that the satellite formerly moving outside is displaced inside and *vice versa.* This behavior has been analysed in detail by Dermott and Murray (1981a,b) and Yoder et al. (1983), and can be expected to occur only for bodies having a very small mass ratio with respect to their primary. The origin of these satellites is not clear: whereas it would be natural to assume that they are collisional fragments from a single parent body (Smith et al. 1982), their relative velocity is much lower than their mutual escape velocity, and this rather supports the idea that the satellites reaccreted independently from a narrow ring of collisional debris. It is also debated whether drag forces, tides and torques by the rings may have affected their dynamical evolution.

Hyperion is located in a 4:3 mean motion resonance with the neighboring, massive Titan, has an orbital eccentricity of about 0.1 and its conjunctions with Titan librate about the apocenter of its orbit, in such a way that close encounters between the two satellites cannot occur (Colombo et al. 1974; Bevilacqua et al.

1980). But as we have seen, its physical properties (in particular, the strongly irregular shape) suggest that Hyperion is the outcome of a catastrophic impact. Thus we can assume that a larger and presumably nearly-spherical proto-Hyperion was disrupted by a collision with a passing body (probably a comet-like object) travelling at a relative speed of the order of 10 km/s. The present Hyperion is just the largest remnant of this breakup, possibly a portion of the core of the parent body, whereas most smaller fragments were ejected at velocities exceeding the the parent's escape velocity, so as to prevent the immediate reaccumulation into a "rubble pile" and to form a narrow ring of debris (Farinella et al. 1983). In most other similar cases, such narrow rings rapidly reaccrete through low-velocity collisions. However, the stable librating orbit of Hyperion (and presumably of its parent body) is surrounded by a limited "island" of stable, regular orbits, and most fragments ejected at speeds higher than ≈ 0.2 km/s did fall out of this stable region of the phase space, into chaotic orbits which underwent repeated close encounters with Titan. Thus Hyperion could not reaccrete from the postimpact ring, and within $\approx 10^3$ yr most fragments collided with Titan, while a small fraction of them might have produced observable craters on the trailing side of the inner satellite Rhea (Farinella et al. 1990). It is possible that the very intense bombardment of Hyperion fragments onto Titan over a short span of time might have affected in a significant way the properties of Titan's atmosphere. In particular, this process could explain the present high abundance of N_2 in the atmosphere, with the nitrogen directly supplied by the impactors and/or released by shock effects in a primordial CH_4- and NH_3-rich mixture (Farinella et al. 1997).

Finally, a case where the "natural selection" process may have occurred only in a very partial way is that of Neptune's satellite system. Its peculiar dynamical features – one massive satellite, Triton, on an inclined retrograde orbit, a small distant satellite, Nereid, on a very eccentric and inclined orbit, and a regular set of small inner moons – indicate that the origin and history of this system must have been different from those of the systems of the other giant planets. A plausible scenario, based on the assumed capture origin of Triton, has been discussed by Farinella et al. (1980) and later developed by Banfield and Murray (1992). According to them, Triton's highly elliptical postcapture orbit forced chaotic perturbations of the orbits of the primordial inner satellites of Neptune. With the exception of Nereid (which ended up on a distant orbit), the resulting large eccentricities led to mutual collisions and destruction of these inner satellites, leaving a disk of debris. Then, the inner satellite system reformed on equatorial orbits after Triton's orbital circularization had been achieved by tides. But it is unlikely that these reformed satellites survived the subsequent flux of cometary impactors. Therefore, today's inner satellites are probably the remnants of precursor ones, destroyed even more recently than Triton's capture.

Acknowledgements. I am grateful to D.R. Davis and F. Marzari for many stimulating discussions. Financial support from the Italian Space Agency and the Human Capital and Mobility Program of the European Union is acknowledged.

References

Anselmo, L., Farinella, P. (1984): Alfvèn drag for satellites orbiting in Jupiter's plasmasphere. Icarus **58**, 182–185.

Asphaug, E., Melosh, H.J. (1993): The Stickney impact of Phobos: A dynamical model. Icarus **101**, 144–164.

Banfield, D., Murray, N. (1992): A dynamical history of the inner Neptunian satellites. Icarus **99**, 390–401.

Bevilacqua, R., Menchi, O., Milani, A., Nobili, A.M., Farinella, P. (1980): Resonances and close approaches: The Titan–Hyperion case. Moon and Planets **22**, 141–152.

Binzel, R.P., Farinella, P., Zappalà, V., Cellino, A. (1989): Asteroid rotation rates: Distributions and statistics. In Binzel R.P., Gehrels T., Matthews M.S. (eds.) *Asteroids II*, Univ. Arizona Press, pp. 416–441.

Capaccioni, F., Cerroni, P., Coradini, M., Farinella, P., Flamini, E., Martelli, G., Paolicchi, P., Smith, P.N., Zappalà, V. (1984): Shapes of asteroids compared with fragments from hypervelocity impact experiments. Nature **308**, 832–834.

Capaccioni, F., Cerroni, P., Coradini, M., Di Martino, M., Farinella, P., Flamini, E., Martelli, G., Paolicchi, P., Smith, P.N., Woodward, A., Zappalà, V. (1986): Asteroidal catastrophic collisions simulated by hypervelocity impact experiments. Icarus **66**, 487–514.

Catullo, V., Zappalà, V., Farinella, P., Paolicchi, P. (1984): Analysis of the shape distribution of asteroids. Astron. Astrophys. **138**, 464–468.

Chapman, C.R., McKinnon, W.B. (1986): Cratering of planetary satellites. In Burns J.A., Matthews M.S. (Eds.) *Satellites*, Univ. Arizona Press, pp. 492–580.

Colombo, G., Franklin, F.A., Shapiro, I.I. (1974): On the formation of the orbit-orbit resonance of Titan and Hyperion. Astron. J. **79**, 61–72.

Colwell, J.E. (1994): The disruption of planetary satellites and the creation of planetary rings. Planet. Space Sci. **42**, 1139–1150.

Colwell, J.E., Esposito, L.W. (1992): Origins of the rings of Uranus and Neptune, 1. Statistics of satellite disruptions. J. Geophys. Res. **97**, 10,227–10,241.

Colwell, J.E., Esposito, L.W. (1993): Origins of the rings of Uranus and Neptune, 2. Initial conditions and ring moon populations. J. Geophys. Res. **98**, 7,387–7,401.

Croft, S.K. (1992): Proteus: Geology, shape, and catastrophic destruction. Icarus **99**, 402–419.

Davis, D.R., Chapman, C.R., Durda, D.D., Farinella, P. and Marzari, F. (1996): The formation and collisional/dynamical evolution of the Ida/Dactyl system as part of the Koronis family. Icarus **120**, 220–230.

Dermott, S.F. (1984): Dynamics of narrow rings. In Greenberg R., Brahic A. (Eds.) *Planetary Rings*, Univ. Arizona Press, pp. 589–637.

Dermott, S.F., Murray, C.D. (1981a): The dynamics of tadpole and horseshoe orbits. I. Theory. Icarus **48**, 1–11.

Dermott, S.F., Murray, C.D. (1981b): The dynamics of tadpole and horseshoe orbits. II. The coorbital satellites of Saturn. Icarus **48**, 12–22.

Esposito, L.W. (1993): Understanding planetary rings. Ann. Rev. Earth Planet. Sci. **21**, 487–523.

Farinella, P. (1987): Small satellites. In Fulchignoni M., Kresák L. (Eds.) *The Evolution of the Small Bodies of the Solar System*, SIF-North Holland Physics Publishers, pp. 276–300.

Farinella, P., Marzari, F. and Matteoli, S. (1997): The disruption of Hyperion and the origin of Titan's atmosphere. Astron. J. **113**, 2312–2316.

Farinella, P., Milani, A., Nobili, A.M., Paolicchi, P., Zappalà, V. (1983): Hyperion: Collisional disruption of a resonant satellite. Icarus **54**, 353–360.

Farinella, P., Milani, A., Nobili, A.M., Paolicchi, P., Zappalà, V. (1985): The shapes and strengths of small icy satellites. In Klinger J., Benest D., Dollfus A., Smoluchowski R. (Eds.) *Ices in the Solar System*, Reidel, pp. 699–710.

Farinella, P., Milani, A., Nobili, A.M. and Valsecchi, G.B. (1980): Some remarks on the capture of Triton and the origin of Pluto. Icarus **44**, 810–812.

Farinella, P., Paolicchi, P., Strom, R.G., Kargel, J.S., Zappalà, V. (1990): The fate of Hyperion's fragments. Icarus **83**, 186–204.

Franklin, F., Lecar, M., Wiesel, W. (1984): Ring particle dynamics in resonances. In Greenberg R., Brahic A. (Eds.) *Planetary Rings*, Univ. Arizona Press, pp. 562–588.

Fujiwara, A. (1991): Stickney–forming impact on Phobos: Crater shape and induced stress distribution. Icarus **89**, 384–391.

Fujiwara, A., Kamimoto, G., Tsukamoto, A. (1978): Expected shape distribution of asteroids obtained by laboratory impact experiments. Nature **272**, 602–603.

Fujiwara, A., Kadono, T., Nakamura, A. (1993): Cratering experiments into curved surfaces and their implication for craters on small satellites. Icarus **105**, 345–350.

Giblin, I., Martelli, G., Smith, P.N., Cellino, A., Di Martino, M., Zappalà, V., Farinella, P., Paolicchi, P. (1994): Field fragmentation of macroscopic targets simulating asteroidal catastrophic collisions. Icarus **110**, 203–224.

Greenberg, R. (1977): Orbit–orbit resonances in the solar system: Varieties and similarities. Vistas Astron. **21**, 209–239.

Harris, A.W. (1984): The origin and evolution of planetary rings. In Greenberg R., Brahic A. (Eds.) *Planetary Rings*, Univ. Arizona Press, pp. 641–659.

Holsapple, K.A. (1994): Catastrophic disruptions and cratering of solar system bodies: A review and new results. Planet. Space Sci. **42**, 1067–1078.

Johnson, T.V., McGetchin, T.R. (1973): Topography on satellite surfaces and the shape of asteroids. Icarus **18**, 612–620.

Lange, M.A., Ahrens, T.J. (1987): Impact experiments in low-temperature ice. Icarus **69**, 506–518.

Lissauer, J.J., Squyres, S.W., Hartmann, W.K. (1988): Bombardment history of the Saturn system. J. Geophys. Res. **93**, 13,776–13,804.

Mignard, F. (1984): Effects of radiation forces on dust particles. In Greenberg R., Brahic A. (Eds.) *Planetary Rings*, Univ. Arizona Press, pp. 333–366.

Plescia, J.B., Boyce, J.M. (1983): Crater numbers and geological histories of Iapetus, Enceladus, Tethys, and Hyperion. Nature **301**, 666–670.

Schenk, P.M. (1989): Crater formation and modification on the icy satellites of Uranus and Saturn: Depth/diameter and central peak occurrence. J. Geophys. Res. **94**, 3813–3832.

Smith, B.A., and the Voyager imaging team (1981): Encounter with Saturn: Voyager 1 imaging science results. Science **212**, 163–191.

Smith, B.A., and the Voyager imaging team (1982): A new look at the Saturn system: The Voyager 2 images. Science **215**, 504–537.

Smith, B.A., and the Voyager imaging team (1986): Voyager 2 in the Uranian system: Imaging science results. Science **233**, 43–64.

Smith, B.A., and the Voyager imaging team (1989): Voyager 2 at Neptune: Imaging science results. Science **246**, 1422–1449.

Stooke, P.J. (1992): The shapes and surface features of Prometheus and Pandora. Earth, Moon, and Planets **62**, 199–221.

Stooke, P.J., Lumsdon, M.P. (1992): The topography of Janus. Earth, Moon, and Planets **62**, 223–237.

Thomas, P.C. (1989): The shapes of small satellites. Icarus **77**, 248–274.

Thomas, P.C., Black, G.J., Nicholson, P.D. (1995): Hyperion: Rotation, shape, and geology from Voyager images. Icarus **117**, 128–148.

Wisdom, J., Peale, S.J., Mignard, F. (1984): The chaotic rotation of Hyperion. Icarus **58**, 137–152.

Yoder, C.F., Colombo, G., Synnott, S.P., Yoder, K.A. (1983): Theory of motion of Saturn's coorbiting satellites. Icarus **53**, 431–443.

II

Physics of Shocks

Concepts of Shock Waves

André Migault

Laboratoire de Combustion et de Détonique (CNRS U.P.R. 9028)
E.N.S.M.A. – B.P. 109 – Site du Futuroscope – Chasseneuil-du-Poitou
F-86960 FUTUROSCOPE Cedex, France

Phénoménologie des ondes de choc

Résumé. Après quelques rappels sur les propriétés d'une onde sonore et la définition d'une onde de choc, on démontre les relations de Rankine-Hugoniot qui traduisent, pour une transformation par choc, les lois classiques de conservation dans un fluide continu ne présentant pas de changement de phase.
Dans les sections 3 et 4, on définit la courbe de Hugoniot dans le plan pression-volume et la polaire de choc dans le plan pression-vitesse particulaire. On examine la position de ces courbes relativement aux courbes de compression et de détente isentropiques, ainsi que la répartition de l'énergie apportée par le choc et l'évolution de l'entropie du milieu soumis au choc.
La section 5 montre quelques résultats expérimentaux obtenus sur des métaux simples et sur quelques corps d'intérêt géologique. L'existence d'une relation linéaire entre la célérité du choc et la vitesse particulaire associée permet de donner une description analytique simple de la courbe de Hugoniot et de la polaire de choc à partir de trois constantes physiques connues dans l'état initial.
Puis on décrit le phénomène d'atténuation hydrodynamique qui est dû à la largeur finie de l'onde de choc et à la croissance de la célérité du son avec la pression.
La section 7 est consacrée à l'étude de la transmission d'un choc entre deux milieux différents, apppliquée dans la section suivante à l'impact d'un projectile plan sur une cible plane semi-infinie, pouvant représenter en première approximation la chute d'une météorite sur une planète.
Enfin, nous abordons le calcul de la température atteinte dans un matériau soumis à un choc, dans les cas basse-pression et haute-pression, de la température résiduelle du matériau quand il s'est détendu, et de la courbe de compression isentropique issue de l'état initial.

Abstract. In this chapter, we develop the classical theory of shock waves in condensed media. A demonstration of Rankine-Hugoniot equations is presented and we study the respective position of Hugoniot, isentropic curves and Rayleigh line in (P,V) plane. Some experimental results are given for metals and geologic materials (relation shock velocity vs. particule velocity). We present the hydrodynamic attenuation (attenuation by release waves). We analyse the transmission of shock between two media and the planar impact against a massive target at the rest; numerical applications are given. In the last part of this paper, we give some indications about the equation of state of condensed materials and the calculus of temperature reached in a shock wave; numerical applications are made.

1 Shock Waves: Definition and Formation

We can define a shock wave as the propagation of a discontinuity of the thermodynamic and mechanical properties of the medium: pressure, volumetric mass, energy, temperature, material velocity.

1.1 Properties of Sonic Waves

Sonic waves are waves which propagate infinitely weak perturbations of various properties around their equilibrium values. They propagate at the sound speed, C_B.

1. these waves propagate perturbations without change of entropy of the medium.
2. the sound speed is given by:

$$C_B = \sqrt{\left(\frac{\partial P}{\partial \rho}\right) S} \tag{1}$$

   P and ρ are the local pressure and volumetric mass, i.e at the place where the wave is at the considered time. S is the entropy. C_B is the local speed of the wave related to the medium at this place; if the medium moves with the material velocity u_p with respect to an absolute frame, the absolute velocity of the wave is $(C_B + u_p)$.
3. the variation of the physical properties and the motion of the medium are submitted to the classical conservation laws of the mass, monentum and energy. For a one dimensional problem, these laws lead to the relation:

$$\rho C_B . \mathrm{d}u_p = \mathrm{d}P \tag{2}$$

   If C_B and $\mathrm{d}u_p$ are in the same orientation, we have a compression wave if $\mathrm{d}P > 0$ and a release wave if $\mathrm{d}P < 0$.
   With the definition (1), (2) gives:

$$\mathrm{d}u_p = C_B \frac{\mathrm{d}\rho}{\rho} \tag{3}$$

4. if the encountered medium is uniform (thermodynamic and mechanical properties independent of spatial coordinates), the speed of the sonic wave is constant and its trajectory in (x, t) plane is a straight line with a slope equal to $C_B + u_p$.
5. the most important propertie is:
   In condensed matter, the sound speed is generally
   an increasing function of pressure or volumetric mass.

1.2 Formation of a Shock Wave

The property (5) is the main point for understanding the formation of a shock wave in condensed matter.

Suppose that we apply an increasing pressure from P_0 to P_1 during a short (but non-zero) time on the surface A of a sample (see Fig. 1). The surface A begins to move and successive sonic waves are emitted from A. One wave propagates in a medium at pressure P and brings it to $P + \mathrm{d}P$; the next wave propagates in a medium at $P + \mathrm{d}P$, so it is faster than the one before, and has tendency to overtake it. Thus, the initial pressure profile is not stable and stiffens: it is a shock wave; its propagation is supersonic relatively to the initial medium: it is a **shock wave**. The main point for such evolution of an applied pressure profile is that the time interval between state 0 and state 1 (P_0 and P_1) must be short; if this time is too long, we can obtain a succession of equilibrium states: a pneumatic does not split if it is drill with a nail but a large tear allows a fast evacuation of air and we hear a sonic bang.

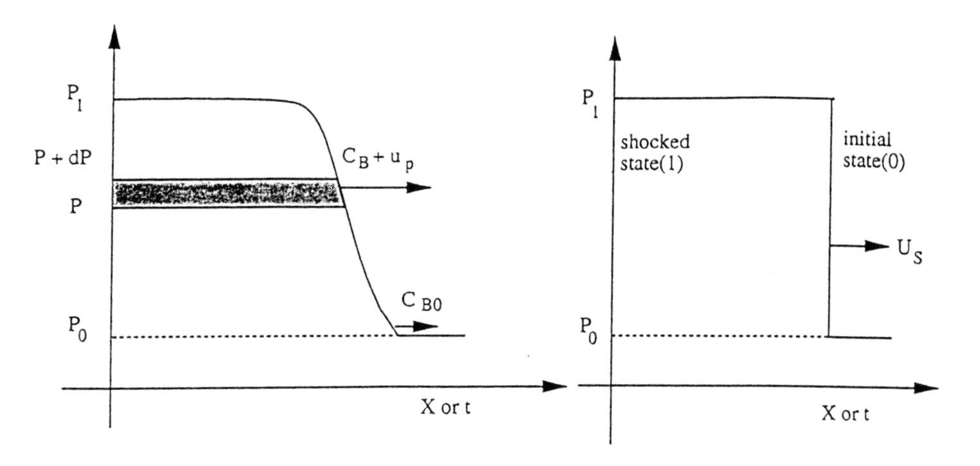

Fig. 1. Formation of a shock wave

Now, we can give a stricter definition of shock wave:

<u>Shock wave is the stable and final form of compressive wave with finite amplitude and short rise time</u>.

We have selected pressure for representing the shock formation. A better representation of the dynamic character of a shock wave is given by the propagation of a material velocity discontinuity; see Eq. (2) or (3).

A discontinuity is a mathematical idea. We can make a detailed analysis of shock wave formation and structure in non-reactive media (Hayes 1960; Landau & Lifchitz 1971). It will involve some calculations with vicosity, heat conduction, variation of sound velocity with pressure, ... Thus, we describe the propagation of a wave with finite amplitude and a thickness different from zero; the velocity of this wave is greater than the one of sound. These characteristics result from competition between anharmonic (non linear) and dissipative effects.

Experimentally, several authors (Doran & Linde 1966; Smith 1958) have given an estimation of temporal thickness of shock fronts in metals: they give 10^{-9} s; it is the present rise time of modern oscilloscopes. For example: in copper, a shock of 70 GPa (1 GPa = 30 kbar = 3. 10^4 bar) propagates at 6000 m/s; the spatial thickness of a shock front is 6 10^{-3} mm; the lattice spacing is 3.6 Å; thus, the shock front covers about 20000 lattice spacings: at the scale of crystal, shock front is macroscopic.

2 Rankine-Hugoniot Equations (Conservation Laws)

Thermodynanic state of material under study is described with only three variables: the pressure P, the volumetric mass ρ and internal energy for unit mass E (or temperature T). P_0, ρ_0 and E_0 (or T_0) are values of these variables in natural state, i.e. unshocked state. This natural state is the state of full rest: particle velocity, u_{p0}, is 0 ahead of the shock front. These three variables are connected with the equation of state (E.O.S.):

$$f(P,\rho,E) = f(P_0,\rho_0,E_0) = 0 \tag{4}$$

A shock wave is characterized by two kinetic variables: U_S, shock velocity and u_p, particle velocity.

These five variables, U_S, u_p, P, ρ and E are connected by three relationships deduced from mechanical conservative laws: conservation of mass, momentum and energy.

2.1 Conservation of Mass, Momentum and Energy in the Laboratory Frame

This demonstration of Rankine-Hugoniot equations is taken from Melosh (1989). Figure 2 shows a piece of material through which a shock wave is travelling; we can see the piece at two different times, t and t' ($t' > t$). We suppose the shock wave is plane, perpendicular to the direction of shock velocity (U_S); thus, the cross-sectional area of the block is constant as the shock moves through it (the cross-sectional area of the block is unity).

At time t, the length of the unshocked region is l_u and the length of the shocked region is l_s. At the time $t' > t$, l_u becomes l'_u and l_s becomes l'_s. Between t and t', the shock wave has covered $U_S(t'-t)$ to the right and shocked end of the piece moving at the particle velocity, u_p, has covered $u_p(t'-t)$ to the right. Thus, the relationships between l and l' are:

$$l'_u = l_u - U_S(t'-t) \tag{5}$$

$$l'_s = l_s + U_S(t'-t) - u_p(t'-t) \tag{6}$$

Mass conservation means that the masses at t and t' are equal. This statement gives, from (5) and (6):

$$\rho(U_S - u_p) = \rho_0 U_S \tag{7}$$

where the volumetric mass of unshocked region is ρ_0 and ρ is the volumetric mass of shocked region.

The momentum and energy of block at the time t' are not equal to the momentum and energy at the time t, because there are external force on the block: pressure P on the left (shocked) end of the block and P_0 on the right (unshocked) end. P is greater than P_0, so a net force $(P - P_0)$ acts toward the right, accelerating material in that direction. The net momentum balance is:

$$\rho l'_s u_p - \rho l_s u_p = (P - P_0)(t' - t) \tag{8}$$

With the aid of (5),(6) and (7), we can obtain:

$$P - P_0 = \rho_0 U_S u_p \tag{9}$$

The work of external forces is the work of P, because only the shocked end (on the left) is moving. Between t and t', this work is $Pu_p(t' - t)$. It is equal to the difference of total energy of the block (E_{tot}) between times t and t'. So, we can write:

$$Pu_p(t' - t) = E'_{tot} - E_{tot} \tag{10}$$

with:

$$\begin{aligned} E_{tot} &= \rho_0 l_u E_0 + \rho l_s E + (1/2)\rho l_s u_p^2 \\ E'_{tot} &= \rho_0 l'_u E_0 + \rho l'_s E + (1/2)\rho l'_s u_p^2 \end{aligned}$$

Now, (10) can be written as:

$$Pu_p = \rho_0 U_S (E - E_0) + (1/2)\rho_0 u_p^2 U_S \tag{11}$$

Some transformations give the three conservation equations (7), (9) and (11) under the following form:

$$u_p = \sqrt{(P - P_0)(V_0 - V)} \tag{12}$$

$$U_S = V_0 \sqrt{\frac{P - P_0}{V_0 - V}} \tag{13}$$

$$E - E_0 = \frac{1}{2}(P + P_0)(V_0 - V) \tag{14}$$

2.2 The Same in the Shock Front Frame

In fluid mechanics, conservation equations are often written in a frame which moves with the shock front. In this frame, the unshocked region (indice 1) has the velocity v_1 relatively to the shock front and the shocked region has the velocity v_2. In this frame, conservation equations are the followings (Hayes 1960; Landau & Lifchitz 1971):

$$\begin{aligned} &\rho_1 v_1 = \rho_2 v_2 \\ &P_1 + \rho_1 v_1^2 = P_2 + \rho_2 v_2^2 \\ &E_1 + \frac{P_1}{\rho_1} + \frac{v_1^2}{2} = E_2 + \frac{P_2}{\rho_2} + \frac{v_2^2}{2} \end{aligned}$$

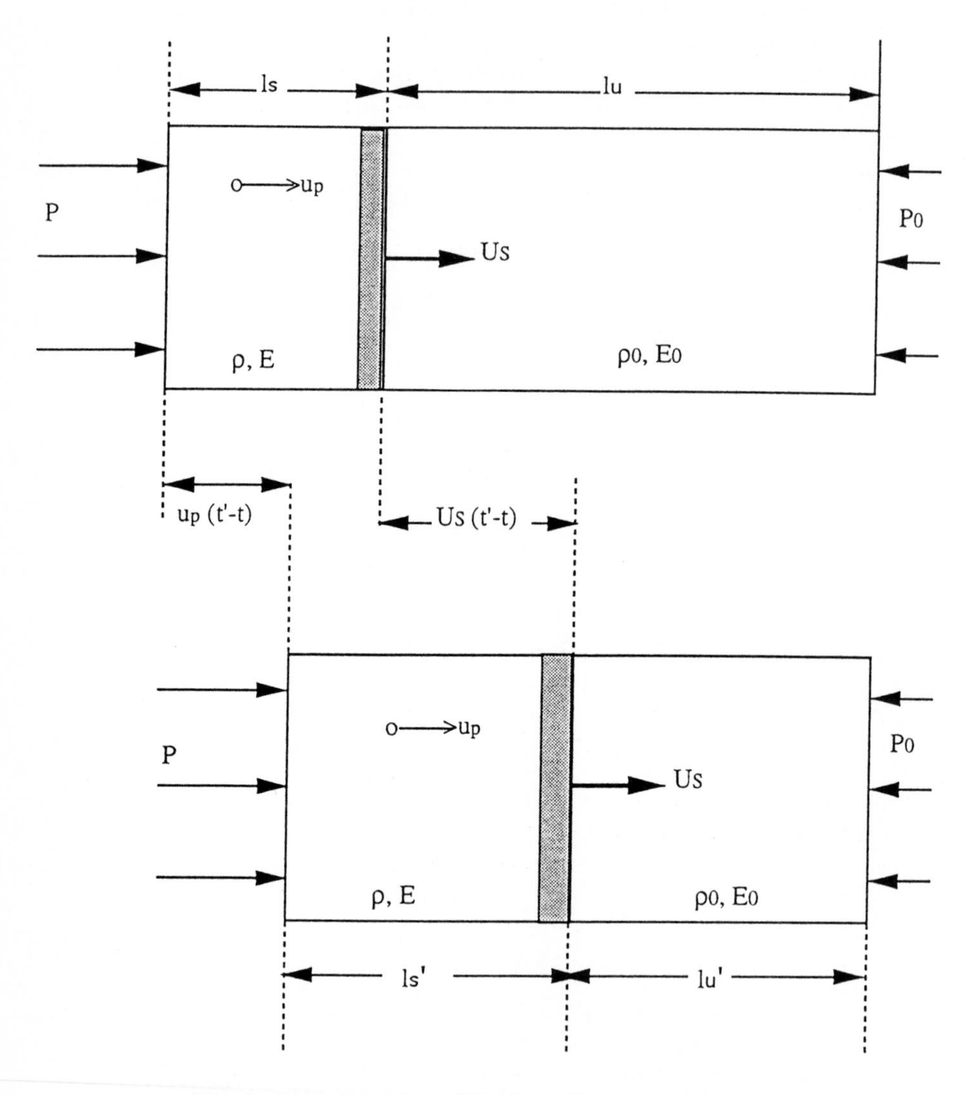

Fig. 2. Demonstration of Rankine-Hugoniot equations

In this frame, we have:

$$U_S = v_1 \qquad and \qquad u_p = v_1 - v_2$$

2.3 The Same in Special Relativity

In following equations, thermodynamic quantities are relating to the unit volume and are evaluated in the proper frame of each element of fluid (i.e. a frame which is comobile with each element of fluid). ω, e and n are the enthalpy, energy and number of particles (after complete annihilation of all pairs particle-antiparticle) per unit volume and $\omega = e + P$.

In the shock front frame, conservation laws are:

$$\frac{n_1 v_1}{\Gamma_1} = \frac{n_2 v_2}{\Gamma_2}$$
$$\frac{\omega_1 v_1}{\Gamma_1^2} = \frac{\omega_2 v_2}{\Gamma_2^2}$$
$$\frac{\omega_1 v_1^2}{c^2 \Gamma_1^2} + P_1 = \frac{\omega_2 v_2^2}{c^2 \Gamma_2^2} + P_2$$

where Γ is the Lorentz factor:

$$\Gamma = \sqrt{1 - \frac{v_2}{c_2}}$$

Shock wave and particle velocities are given by:

$$u_p = \frac{v_1 - v_2}{1 - \frac{v_1 v_2}{c^2}} \qquad and \qquad U_S = v_1$$

3 Hugoniot Curve and Its Representation in the (P, u_p) Plane

The medium state behind the shock is characterized by:
- 3 thermodynamical variables: P, ρ and E;
- 2 kinetic variables: U_S and u_p.

Between these 5 variables, we have 4 relationships:
- 3 conservative laws (see Eqs. (12), (13) and (14));
- 1 equation of state (see Eq. 4).

Thus, we have relationship between the variables 2 by 2. The two most important relationships are:

- $P_H(V)$ which defines the "Hugoniot curve";
- $P_H(u_p)$ which defines another curve in (P, u_p) plane (in french, we call this curve: "polaire de choc"). This curve is very important in study of shock transmission between two media.

Figure 3 shows the Hugoniot curve in (P, V) plane. 0 represents the initial state (P_0, V_0, E_0) and 1 is the final state (P_H, V, E_H) (where V is the independent variable) which is obtainted through a single shock.

Curves S_0 and S_1 are isentropic curves relating to state 0 and state 1. We will see in the following how they are located relatively to the Hugoniot curve in (P, V) plane. Equations (12)-(14) show that,

for one specified initial state, we have one Hugoniot curve and one only.
This curve is not a thermodynamic curve but a geometrical locus:
conservative laws give no way to connect initial state to final state.

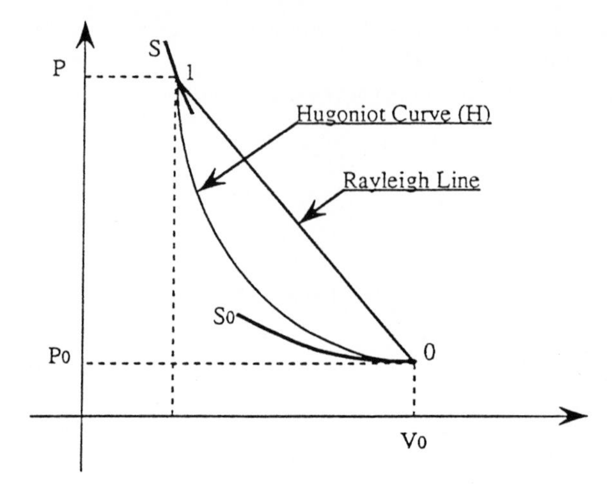

Fig. 3. Hugoniot curve and Rayleigh line connecting initial and final state

Straight line between 0 and 1 is the Rayleigh line. It is not a thermodynamic curve. We will see that entropy has a maximum value on this line, between 0 and 1. The slope of this line is proportional to U_S^2. Figure 4 shows the representation of Hugoniot curve in (P, u_p) plane. The slope of the straight line between initial state 0 and final state 1 is $\rho_0 U_S$: this quantity is called: shock impedance of the medium.

4 Comments on Conservative Laws and Relative Position of Hugoniot and Isentropic Curves in the (P, V) Plane

4.1 Comments on Conservation Laws

In (11) or (14), we substitute U_S and u_p to P and V. We obtain:

$$E - E_0 = \frac{1}{2}u_p^2 + P_0 V_0 \frac{u_p}{U_S} \tag{15}$$

Usually, P_0 is the atmospheric pressure, about 1 bar; values of V_0 are included between 0.05 and 1 cm^3/g for solids and liquids; $P > P_0$, then $u_p/U_S < 1$. Thus, the second term of (15) ranges about 0.01 J/g. On the other hand, for an ordinary shock in dense medium, we have $u_p \approx 1mm/\mu s$ and the first term of (15) is about 500 J/g; we conclude that the second term of (15) may be neglected and we rewrite (15) as:

$$E - E_0 \approx \frac{1}{2}u_p^2 \tag{16}$$

Equation (16) shows that total energy given by the shock is equally distributed between kinetic energy and internal energy $(E - E_0)$. Thus, high temperatures

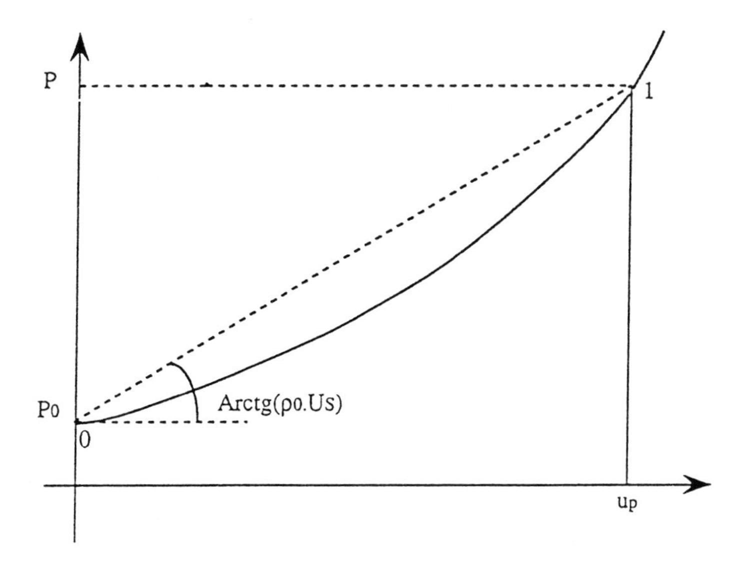

Fig. 4. Hugoniot in (P, u_p) plane

are associated with high dynamic pressures: in shock waves, we can observe crystalline phase change, melting, ...

For a weak shock, we have $P \approx P_0$ and (14) gives:

$$E - E_0 \approx P(V_0 - V) \tag{17}$$

This equation can be compared with the thermodynamic relation:

$$\delta E = -P\delta V + T\delta S \qquad \text{with} \quad \delta S = 0 \tag{18}$$

Equations (17) and (18) show there is continuity between weak shock and isentropic compression.

With Rankine-Hugoniot relation (14) and the thermodynamic relation (18), we can demonstrate that:

1. infinitely weak shock is isentropic;
2. Hugoniot curve (H) and isentropic curve crossing initial state (S_0) have a second order contact at this point in (P,V) plane;
3. the entropy variation is of the third order in volume variation:

$$S - S_0 \approx -\frac{1}{12T_0}(V - V_0)^3\left(\frac{d^2P}{dV^2}\right)_{H,V_0} + \text{term.}(V - V_0)^4 \; ; \tag{19}$$

4. entropy and temperature are monotonic increasing functions of pressure along Hugoniot curve.

4.2 Position of Hugoniot Curve (*H*) Relatively to Isentropic Curves in the (*P*, *V*) Plane

We recall some properties of isentropic curves.

1) for each (P,V) state there is one isentropic curve and only one. It is a result of existing equation of state.

Thus, two any isentropic curves have no common point.

2) from (1), we have:

$$(\frac{dP}{dV})_S = -\frac{C_B^2}{V^2}$$

C_B is an increasing function of $1/V$ and concavity of isentropic curves is always directed upwards.

3) temperature is an increasing function of pressure during an isentropic compression. In this case, we have:

$$(\frac{\partial P}{\partial S})_V = -(\frac{\partial T}{\partial V})_S > 0 \tag{20}$$

If volume V is imposed, pressure increases with entropy.

Thus, if $S > S_0$, isentropic curve with S is always completely above isentropic curve with S_0 in (P,V) plane.

We consider Fig. 5 in (P,V) plane. 1 is a shocked state connected to initial state 0 through a single shock. S_1 and S_0 are isentropic curves sprung from states 1 and 0. Their position and form result from the first and third above propositions. In 4.1., we have said entropy increases along Hugoniot curve; thus, 0 and 1 are the single common points of Hugoniot and isentropic S_0 and S_1.

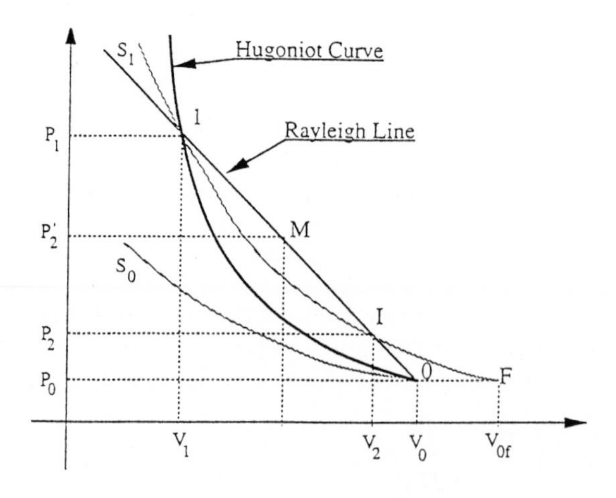

Fig. 5. Position of Hugoniot curve relatively to isentropic curve in (P,V) plane

For any transformation, we have:

$$T\,dS = dE + P\,dV$$

Integrating this relation along straigth line 01, we obtain:

$$\int_0^1 T\,dS = E_1 - E_0 + \int_0^1 P\,dV) = \frac{1}{2}(P_1 + P_0)(V_0 - V_1) + \int_0^1 P\,dV) \quad (21)$$

The first term on the right is the area of trapezium (V_1,1,0,V_0) and the second is just this area with opposite sign:

$$\int_0^1 T\,dS = 0$$

T is always positive and S is an increasing function of P: thus, differential dS must be zero at one point between 0 and 1 and at this point, S is maximum or minimum.

We have the two following possibilities (see fig. 6, drawn in (P,S) plane): if S has a maximum between 0 and 1 along the straigth line 01, its variations are described by the (b) curve. It is the (a) curve in the case of minimum of S. The (H) curve represents the variations of entropy between 0 and 1 along the Hugoniot curve. If S has a minimum along straigth line 01, (a) curve shows that entropy along line 01 crosses over the value S_0 for $P = P_3$ between P_0 and P_1; thus, isentropic curve S_0 re-cuts line 01 and the isentropic curve S_1: that is impossible owing to above propositions 1 and 2. We inferred that entropy along 01 line has a maximum and reaches the value S_1 at one point I between 0 and 1; the maximum value of entropy along 01 is S_1', for pressure $P_2' > P_2$ at M point between 0 and 1. The portion $1I$ of isentropic S_1 is between Hugoniot curve and the Rayleigh line and the portion IF is above the Rayleigh line.

At points 0 and 1 we have the following relationships:

$$\text{slope } 01 < \text{slope of } S_1 \text{ at } 1 \quad (22)$$

$$\text{slope } 01 > \text{slope of } S_0 \text{ at } 0 \quad (23)$$

Few mathematics give using eqn.(7):

$$C_{B1} + u_P > U_S \quad (24)$$

and:

$$U_S > C_{B0} \quad (25)$$

Relationships (24) and (25) are written in laboratory frame (initial medium where the shock is moving). They show that a shock wave is subsonic relatively to shocked medium and supersonic relatively to the initial medium. Inequality (24) shows that a shock wave will be overtake by sonic perturbations which are emitted in the shocked medium; this result explain hydrodynamic attenuation (cf. Sect. 6).

Some calculations give approximate relationship:

$$U_S \approx \frac{1}{2}[(C_{B1} + u_p) + C_{B0}] \quad (26)$$

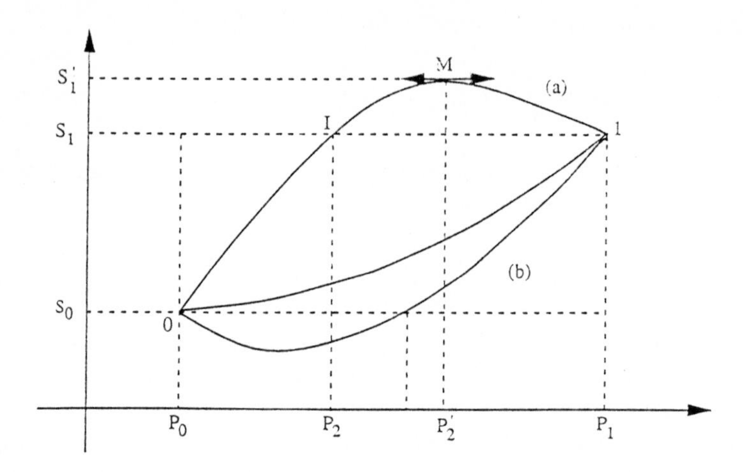

Fig. 6. The two possibilities for the entropy between state 0 and 1 connected by shock wave

The shock wave velocity is the arithmetical mean of the sound velocities in shocked and unshocked mediums (all these velocities are evaluated in the same frame). From shock wave measurements, we obtain U_S and u_p; static measurements give C_{B0}; thus, (26) gives a rapid estimation of sound velocity, C_{B1}, under pressure .

5 Experimental Results

5.1 Some Experimental Results on Simple Metals

Among the 3 thermodynamic variables and the 2 kinematic variables which are in the conservation laws, only the kinetic variables U_S and u_p can be otained from reliable measurements (some attempts are made to measure directly shock pressure (see general litterature). We obtain pairs of values (U_S,u_p). Figure 7 shows results obtained for 7 metals. For each metal we have plotted U_S vs. u_p.

With the aid of conservation laws for mass and moment, we can calculate P_H and V/V_0 and draw $P_H(V)$ and $P_H(u_p)$. Figure 8 shows $P_H(u_p)$ for the 7 metals under examination.

5.2 Relation Between U_S (Shock Wave Velocity) and u_p (Particular Velocity)

For a large number of materials, particularly metals, experiments show a linear relationship between U_S and u_p:

$$U_S = C + S.u_p \tag{27}$$

The values of coefficient C are very near of the values of sound velocity at zero pressure (C_{B0}). Equation (27) shows that C is the value of U_S for $u_p = 0$ and we have seen in Sect. 4 a weak shock is an isentropic compression which propagates at sound velocity. Figure 9 shows the correlation which exists between C and C_{B0}.

Values of ρ_0, C, S, C_{B0} and γ_0 for the 7 metals of Figs. 7 and 8 are listed in Table 1 (where γ_0 is the Gruneisen coefficient of material. We define it in Sect. 9).

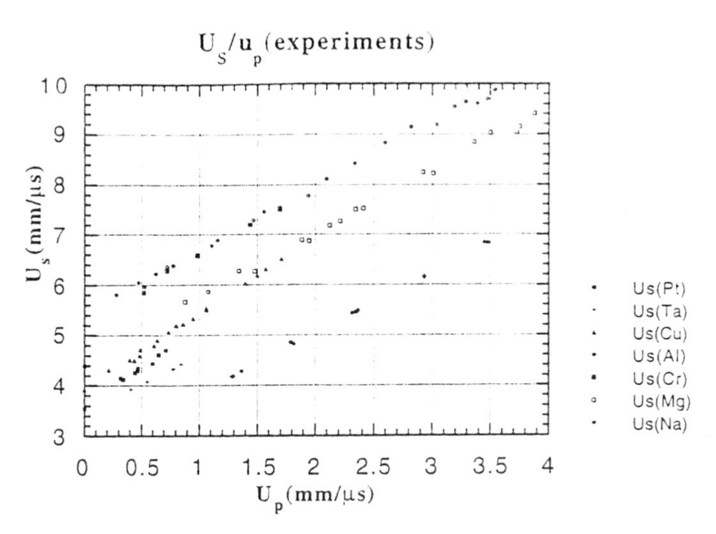

Fig. 7. U_S vs. u_p: experimental results for seven metals

Table 1.. Values of ρ_0, C, S, C_{B0} and γ_0 for the 7 metals of Figs. 7 and 8

	ρ_0 (g/cm^3)	C (mm/μs)	S (-)	γ_0 (-)	C_{B0} (mm/μs)
Pt	21.449	3.68	1.46	2.90	3.538
Ta	16.656	3.43	1.19	1.80	3.388
Cu	8.924	3.91	1.51	2.00	3.927
Al(2024)	2.784	5.37	1.29	2.00	5.209
Cr	7.119	5.20	1.43	1.50	4.742
Mg	1.740	4.50	1.26	1.60	4.440
Na	0.968	2.58	1.24	1.30	2.350

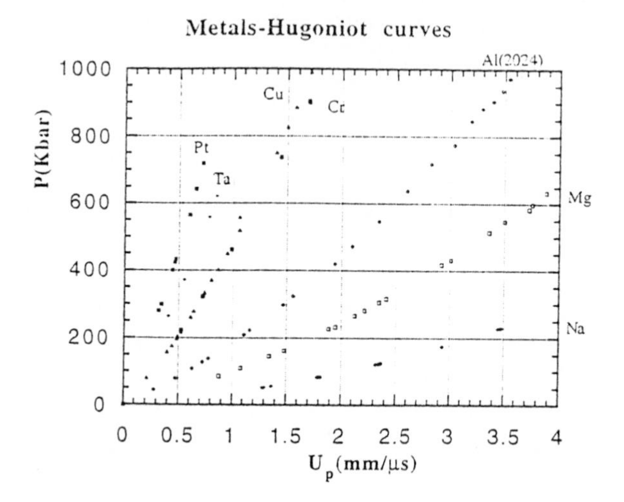

Fig. 8. Hugoniot curves in (P, u_p) plane for seven metals

With C, S and ρ_0 and conservation equations (7) and (9), we can obtain an analytical form for $P_H(V)$ and $P_H(u_p)$. These forms are the following:

$$U_S = C + S\ u_P$$

$$P_H(u_p) = \rho_0(C + S\ u_p)u_p \tag{28}$$

$$P_H(V) = \rho_0 C_2 \frac{1 - V/V_0}{[1 - S(1 - V/V_0)]^2} \tag{29}$$

For Eqs. (27)-(29), following units are employed:
- velocities in mm/μs,
- volumetric mass in g/cm^3,
- pressure in GPa.

Equation (29) displays a double pole ($X_P = V_P/V_0 = 1 - 1/S$) which has no physical meaning: so, we can not extrapolate (29) too far ahead. In strict accordance, we can use (27)-(29) only in the vicinity of experimental domain where (27) has been established. With very simple equation of state, such as perfect gas, we can deduce from (7) and (9) an exact relationship between U_S and u_p which gives correct limits when P_H goes to zero and to infinite.

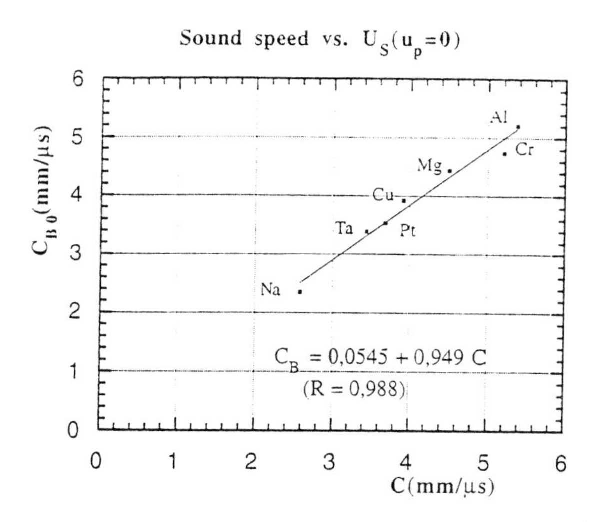

Fig. 9. Correlation between sound speed velocity and coefficient C of Eq. (27)

Figure 10 shows $P_H(u_p)$ drawn with Eq. (28).

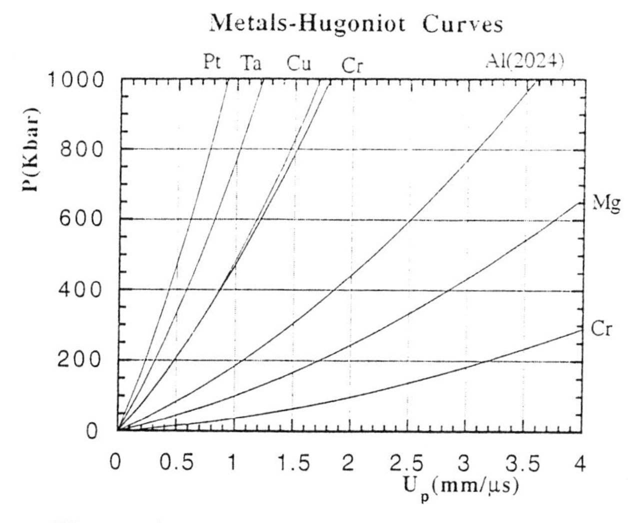

Fig. 10. Analytical Hugoniot curves in (P, u_p) plane

5.3 Some Results on Geological Materials

Figure 11 shows curves U_S vs. u_p for three geologic materials: calcite ($CaCO_3$), ceramic quartz (SiO_2) and forsterite (Mg_2SiO_4).

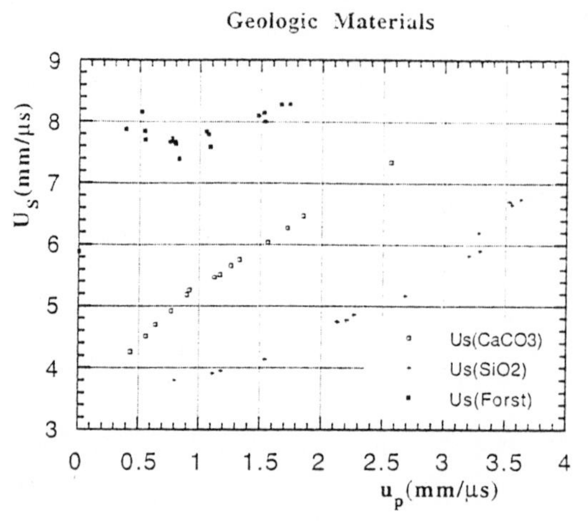

Fig. 11. U_S vs. u_p for three geologic materials

Table 2 gives ρ_0, C, S, γ_0 and C_{B0} for these materials. As we can see on Fig. 11, the curves U_S vs. u_p are not straight line, but curves $P_H(u_p)$ are almost parabolic. We have used these curves for computation of C and S. Results are given in Table 2.

Table 2. ρ_0, C, S, γ_0 and C_{B0} for three geologic materials

	ρ_0 (g/cm^3)	C (mm/μs)	S (-)	γ_0 (-)	C_{B0} (mm/μs)
$CaCO_3$	2.703	4.083	1.293	1.18	
SiO_2	2.145	2.344	1.085	0.90	
Mg_2SiO_4	3.201	6.818	0.766	1.31	5.888

Figures 12 and 13 show $P_H(u_p)$ and $P_H(V)$ obtained from experimental points U_S/u_p (Fig. 11) and conservation equations.

5.4 Numerical Compendiums

The main compendiums of numerical results on shock waves which we use are listed below:

Marsh S.P., LASL – Shock Hugoniot Data, University of California Press, Berkeley, 1980

Compendium of shock wave data, University of California at Livermore – UCRL 50108, June 1966

McQueen R.G. et al., "The Equation of State of Solids from Shock Wave Studies", In Kinslow R.(ed.) *High-Velocity Impact Phenomena*, Academic Press New York and London, 1970

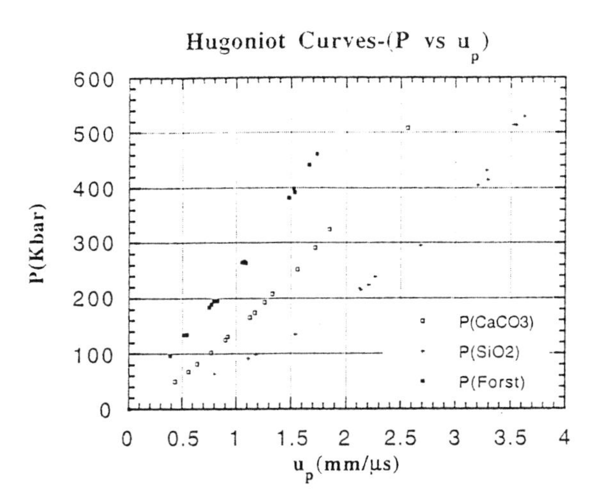

Fig. 12. Hugoniot curves in (P, u_p) plane for the three geologic materials

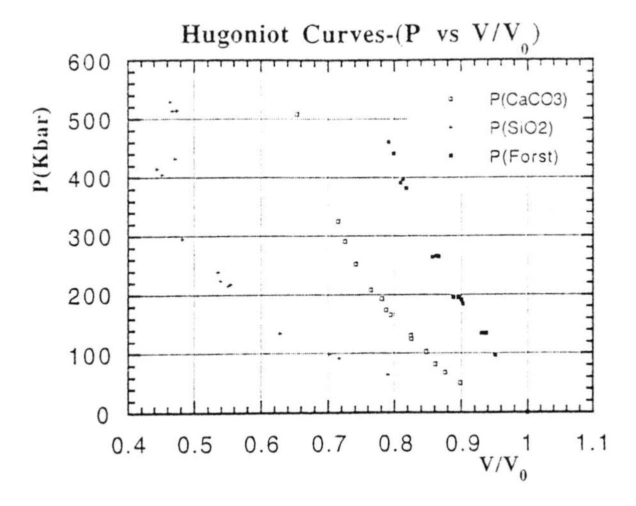

Fig. 13. Hugoniot curves in (P, V) plane for the three geologic materials

6 Hydrodynamic Attenuation

We consider a square pressure pulse travelling through a sample of material. We know the characteristics and the Hugoniot curve of this material. Induced pressure is P_1 and this pressure is held during time τ. During this time, the face subjected to shock moves with the particular velocity u_p; beyond this time, this face stops and a flow of release waves is emited from it. These wavess propagate at the sound speed in the compressed medium in the direction of the initial shock. In the laboratory frame, the shock velocity is U_S and particular velocity is u_p and the velocity of release waves is $C_B + u_p$, where C_B is the sound velocity at pressure P; this velocity varies between C_{B1} (for $P = P_1$) and C_{B0} (for $P = P_0$, pressure of initial state).

$$C_{B0} < C_B < C_{B1} \tag{30}$$

We have seen that shock wave is subsonic relatively to shocked medium; thus:

$$C_{B1} + u_p > U_S \tag{31}$$

It is obvious that the first release waves will overtake and damp the shock: that is the hydrodynamic decay. Meanwhile, as the shock is supersonic relatively to initial medium, the foot of the profile becomes larger. This change in pressure profile is shown on Fig. 14.

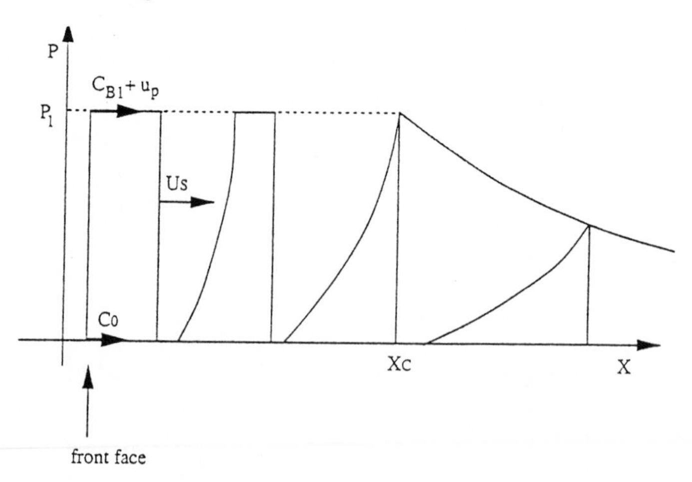

Fig. 14. Hydrodynamic attenuation: evolution of pressure profile during propagation

The damping of the pressure pulse begins when the first release wave overtakes the shockfront. This takes place at the time T_C, when shock wave has covered interval X_C. Figure 15 shows that:

$$X_C = U_S T_C = u_p \tau + (C_{B1} + u_p)(T_C - \tau)$$

we deduce:

$$T_C = \frac{\tau C_{B1}}{C_{B1} + u_p - U_S} \qquad X_C = U_S T_C \tag{32}$$

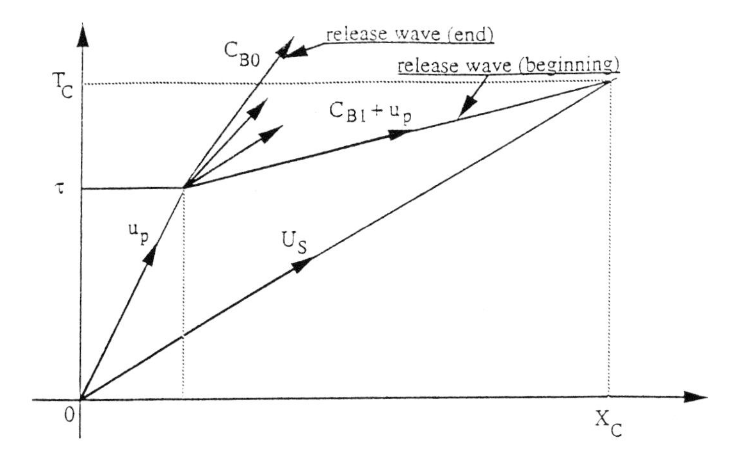

Fig. 15. Hydrodynamic attenuation: wave trajectories in (time-space) plane

6.1 Hydrodynamic Decay of Pressure Pulses

From this time T_C, the amplitude of shock front decreases from P_1 to P_0: this is an adiabatic decay and the pressure pulse becomes triangular. We give below a method to compute this decay (Fig. 16).

Between t and $t + \Delta t$, shock front has covered $\Delta X = U_S \Delta t$ and point A has covered $(\Delta s + \Delta X)$. We have:

$$\Delta t = \frac{\Delta X}{U_S} = \frac{\Delta s + \Delta X}{C_B + u_p} \quad \text{and} \quad \Delta s = \Delta X (\frac{C_B + u_p}{U_S} - 1) \tag{33}$$

The slope of the pressure profile behind the shock front (gradient) is $\partial P / \partial X \approx -\Delta P / \Delta s$. From (33), we have:

$$\Delta P \approx -\Delta s (\frac{\partial P}{\partial X})_1 = -\Delta X (\frac{C_B + u_p}{U_S} - 1)_1 (\frac{\partial P}{\partial X})_1 \tag{34}$$

(34) gives the decay of pressure between (1) and (2) (see Fig. 16). The lower index "1" in (34) means we must compute expressions at the point (1).

During Δt, the back of the wave has covered $C_{B0} \Delta t$ and the head has covered $U_S \Delta t$. Thus, at time $t + \Delta t$, the width of the pulse is:

$$(DE) = (BC) + \Delta X (\frac{U_S - C_{B0}}{U_S})_1 \tag{35}$$

States (1) and (2) are connected by (34) and (35) and we can compute states by degrees.

The change in sound velocity C_B with pressure can be computed with the aid of the Hugoniot curve and a model of equation of state. With the Mie-Gruneisen equation of state (Sect. 9), we have:

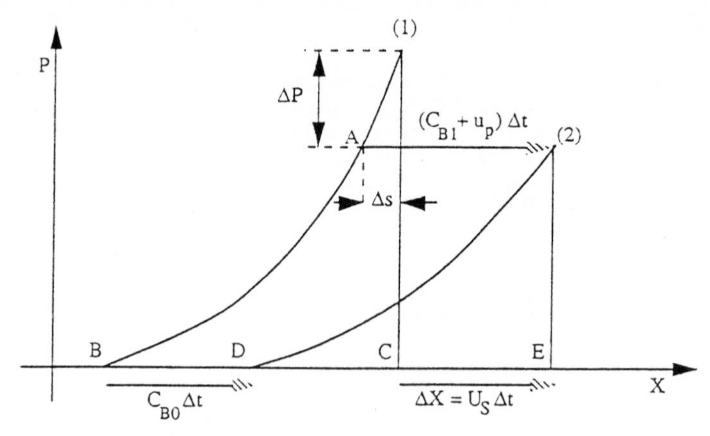

Fig. 16. Hydrodynamic attenuation: calculus

$$C_B^2 = C_H^2[1 - \frac{\gamma}{V}(V0 - V)] + \frac{\gamma}{2V}V^2(P_H - P_0) \qquad (36)$$

$$C_H^2 = -V^2\frac{dP_H}{dV}$$

In Boustie (1991) and Cottet (1985), the reader will find more informations on these calculations and some results.

It is obvious that hydrodynamic process of shock wave decay is not the only cause for shock wave decay; other causes of decay are:

– viscosity and thermal conduction which are not taken into account in the conservation laws.

– bidimensional effects which result from finite dimension of the wave propagating through the sample.

7 Shock Transmission Between Two Media

We are interested in the following situation (see Fig. 17): two samples of different materials are in contact and a shock wave propagates from (A) to (B). The two medias (A) and (B) are initially at rest (particular velocities are zero and pressure is the atmospheric pressure, P_0).

Shock wave brings the medium (A) to the state (1): pressure P_1 and particular velocity u_{p1}. During its propagation, the shock wave reaches the boundary $(A)/(B)$. At the crossing of this boundary, there is:

– shock wave transmission in (B)

– reflexion of a shock or a release wave in (A);

After the transmission, on both sides of the boundary, there is:

– pressure balance

– particular velocity balance.

There are conditions of mechanical equilibrium between vectorial quantities. On the other hand, thermodynamic quantities such as ρ, P, T and S (which are numerical quantities) may have different values on both sides of the boundary:

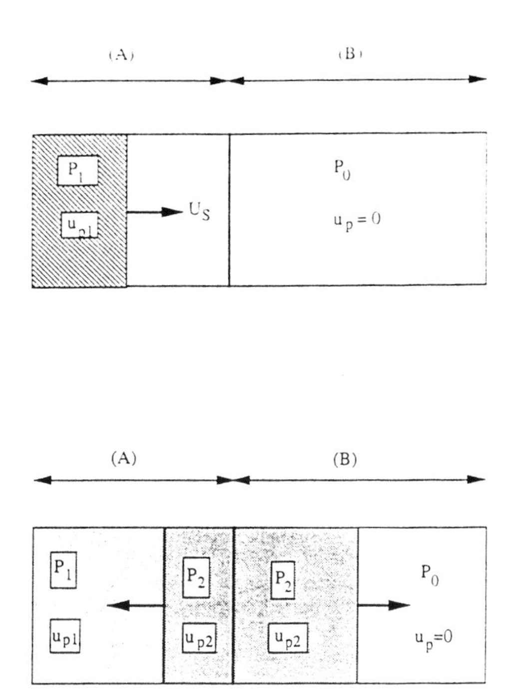

Fig. 17. Shock transmission between (A) and (B): **a** before contact, **b** after contact

for these, the boundary is a surface of discontinuity. We know state (1) induced by shock in medium (A) and we will compute state (2) induced by the shock in medium (B) taking into account the conditions of mechanical equilibrium. For this, the process is analysed in the (P, u_p) plane with the shock and release state curves characteristic of (A) and (B) (see Fig. 18).

Incident shock raises medium (A) in the state (1): P_1, u_{p1}. This state is represented by the point A_1 on the hugoniot curve of (A). Medium (B) is initially at rest and it goes towards state B_2 on hugoniot curve of (B). The co-ordinates of B_2 are (P_2, u_{p2}). Hugoniot curves of (A) and (B) are different, so A_1 and B_2 are separate. For mechanical equilibrium, it is wanted that another wave modifies the state of medium (A) and raises the medium from A_1, on hugoniot curve of (A), to B_2, on hugoniot curve of (B). This is the reason why there is a reflected wave at the boundary $(A)/(B)$ propagating through (A) from boundary $(A)/(B)$. This wave propagates backwards in (A), from $(A)/(B)$, (in the opposite sense of incident shock wave) and it induces the jumps $(P_2 - P_1)$ and $(u_{p2} - u_{p1})$. There are two situations:

- $P_2 > P_1$ reflected wave is a shock wave
- $P_2 < P_1$ reflected wave is a release wave

7.1 Reflected Shock Waves

The jump $(u_{p2} - u_{p1})$ induced by reflected wave has the same sign as the velocity of the wave, so the sign opposite from $\mathbf{u}_{p1}$: medium (A) slows down. Thus, we have:

$$(\mathbf{u}_{p2} - \mathbf{u}_{p1}).\mathbf{u}_{p1} < \mathbf{0} \qquad \text{so} \qquad \mathbf{u}_{p2}.\mathbf{u}_{p1} < \mathbf{u}_{p1}^2$$

We conclude that:

$$\textbf{For } \mathbf{P}_2 > \mathbf{P}_1 \quad \Longrightarrow \quad \mathbf{u}_{p2} < \mathbf{u}_{p1}$$

The locus of possible states for medium (A), shocked from A_1, is the hugoniot curve of 2nd order, H_1' with the pole A_1. For this shock from A_1, conservative laws (cf. section 2) can be used if we take into account that the matter in state A_1 moves with the particular velocity u_{p1} relatively to the initial medium at pressure P_0.

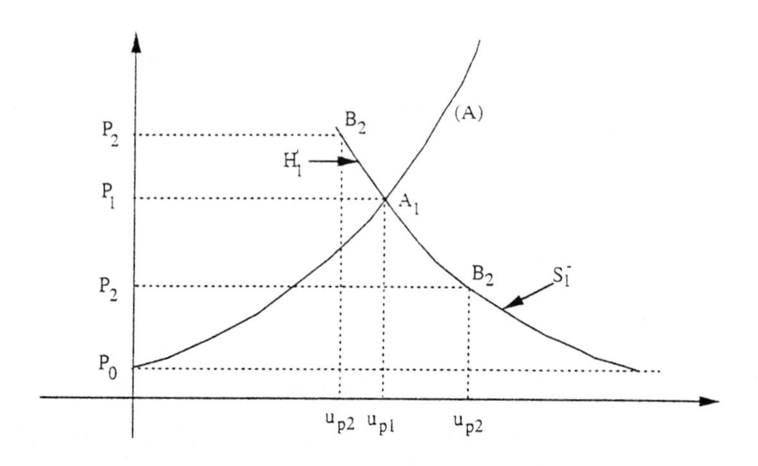

Fig. 18. Shock transmission between (A) and (B) in (P, u_p) plane

7.2 Reflected Release Waves

The release wave is a set of sonic waves which propagates in the opposite sense of incident shock; elementary variations of particular velocity $\mathbf{du}_p$ that are induced have the same sign as $\mathbf{u}_{p1}$: medium (A) speeds up. We can write:

$\mathbf{u}_{p1}.\mathbf{du}_p > 0$

$\int_{\mathbf{u}_{p1}}^{\mathbf{u}_{p2}} \mathbf{u}_{p1}.\mathbf{du}_p = \mathbf{u}_{p2}.\mathbf{u}_{p1} - \mathbf{u}_{p2}^2 > 0$

We conclude:

$$\textbf{For } \mathbf{P}_2 < \mathbf{P}_1 \quad \Longrightarrow \quad \mathbf{u}_{p2} > \mathbf{u}_{p1}$$

The locus of possible state for (A), released from A_1, is S_{1-}, reflected isentropic release curve (see Fig. 18). Along this curve, u_p increases and P decreases; in this case, equation (2) may be written:

$$du_p = -C_B \frac{\mathrm{d}\rho}{\rho}$$

$$u_p = u_{p1} - \int_{\rho_1}^{\rho} C_B \frac{\mathrm{d}\rho}{\rho}$$

This integral can be computed and gives the curve S_{1^-} in the (P,u_p) plane if we know the state A_1.

In conclusion: the locus of possible states taken by (A) in reflected wave at the boundary (A)/(B) is made of the two curves H_1' and S_{1^-} bounded at point A_1. The state of (A) taken after transit of reflected wave from the boundary is one of the two points B_2.

7.3 Equilibrium State at the Interface

The nature (shock or release wave) of the reflected wave depends on the relative position of Hugoniot curves of (A) and (B) in (P,u_p) plane. In both cases, the state at the boundary is located at the crossing of Hugoniot curve of (B) with the locus of possible reflected states of (A) from A_1: this locus is H_1' or S_{1^-}. If Hugoniot curve of (B) is above that of (A), the reflected wave is obviously a shock wave: the state is B_2, located at the crossing of Hugoniot curve of (B) with H_1'. If Hugoniot curve of (B) is below that of (A), the reflected wave is a release one: the state is B_2, located at the crossing of Hugoniot curve of (B) with S_{1^-}.

7.4 Approximation on Curves H_1' and S_{1^-}

Numerical computations in (P,u_p) show that

<u>H_1' and S_{1^-} are close to H_0',</u>
<u>symetrical curve of H_0 relatively to the straight line $u_p = u_{p1}$.</u>
<u>H_0 is the Hugoniot curve relatively to initial state.</u>

<u>This approximation is valid only for condensed matter</u>
<u>and for pressure P_1 not too high.</u>

8 Planar Impact of Projectile on Semi-infinite and Massive Target at Rest

We examine the following situation:

A projectile (A) with parallel faces, thickness "e" and mass "m" is launched with velocity V on a semi-infinite and massive target (B) whose impacted face is parallel to those of the projectile. These faces are orthogonal to the velocity V. The movement of projectile is from left to right.

After impact of (A) against (B), a shock is induced into (B) and a set of shock or release waves takes place in the system (A) + (B). It is this set of waves we want to study. We suppose that all waves are plane and orthogonal to V.

8.1 Chronology of Events

For this, see Fig. 19, drawn in (x,t) plane. This frame is bounded with initial position of target. Velocities are measured in this frame.

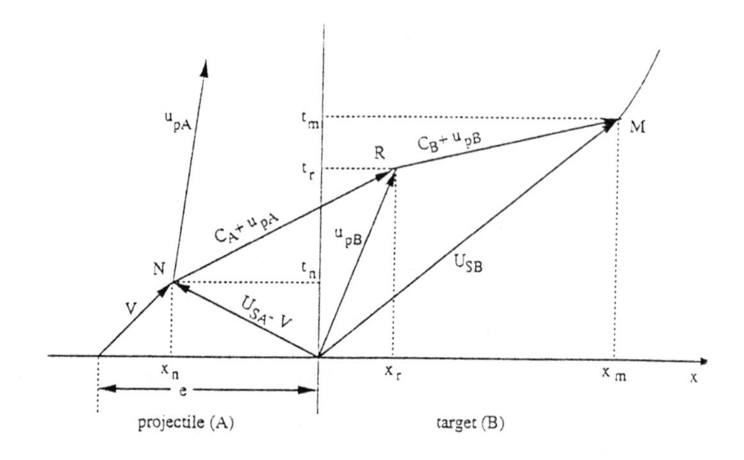

Fig. 19. Planar impact: wave trajectories in (time-space) plane

1)- at the impact of (A) against (B), a shock is induced in (B): pressure P_B, velocity U_{SB} and particular velocity, u_{pB}. The instant of impact is taken as temporal beginning of different events.
2)- the impacted face of target moves forward with velocity u_{pB} and
3)- a shock wave goes back into projectile (A) with velocity $(U_{SA}-V)$. Associated particular velocity is $u_{pA} = V - u_{pB}$.
4)- during this time, the rear face of projectile moves forward with velocity V.
5)- at instant t_n, the wave 3), which goes back into projectile, meets the rear face (point N of Fig. 19).
6)- after this time,

– the rear face of projectile slows down and moves forward at the velocity u_{pA}

– a set of release waves, coming from the rear face of projectile (A), propagates from left to right and runs after the shock (P_B,U_{SB}). We assimilate this set to a simple wave with celerity $(C_A + u_{pA})$ (where C_A is the celerity of sound in medium A at pressure P. If $P = P_0$, this celerity is equal to C_{0A}). The trajectory of this wave is NR (see Fig. 19).

7)- at R, this release wave meets the boundary $(A)/(B)$ at the time t_r; it is transmitted along the trajectory RM with celerity $(C_B + u_{pB})$. This refraction is followed up with a sonic wave which goes back into the projectile.
8)- at point M, the release wave sprung from point R, has overtaken the incident shock in target and begun to damp it: this is the hydrodynamic decay (see section 6); it begins at t_m.

The theory is described below; events have the same number in Sects. 8.1. and 8.2.

8.2 Theory

1)- induced state: we use the results of Sect. 7. The induced state in target (B) is located at the crossing of Hugoniot curve of (B) and reflected Hugoniot of (A); this reflected Hugoniot runs through the point of co-ordinates $u_p = V$ and $P = P_0$; this point represents the mechanical state of projectile (A) at the time of impact. In (P,u_p) plane, this reflected Hugoniot of medium (A) is symmetrical of the Hugoniot curve about the straight line $u_p = V/2$ (see Sect. 7.4). These curves are drawn on Fig. 20.

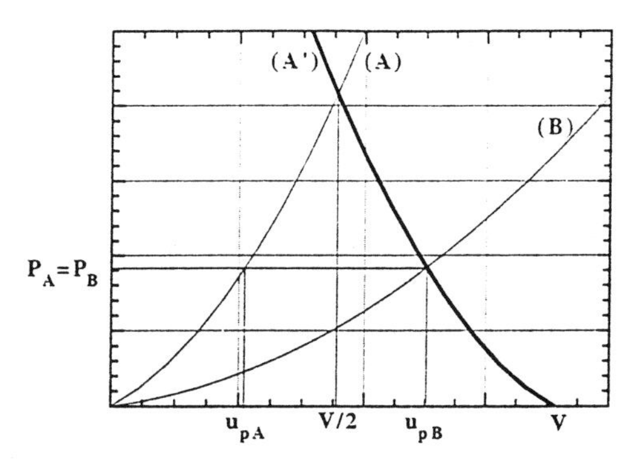

Fig. 20. Determination of state in planar impact in (P, u_p) plane

If we assume that the two media, (A) and (B), have a linear relationship between U_S and u_p, the particular velocity u_{pB} of the shock induced into (B) is given by the following equation:

$$\rho_{0A}[A_A + S_A(V - u_p)](V - u_p) = \rho_{0B}(A_B + S_B u_p)u_p \tag{37}$$

Equation (37) may be written:

$$Qu_p^2 + Tu_p + X = 0 \tag{38}$$

with the following notations:

$$\begin{aligned} Q &= S_B - (\frac{\rho_{0A}}{\rho_{0B}})S_A \\ T &= A_B + (\frac{\rho_{0A}}{\rho_{0B}})(A_A + 2S_A V) \\ X &= -(\frac{\rho_{0A}}{\rho_{0B}})(A_A + S_A V)V \end{aligned}$$

The quantities A, S and ρ_0 are defined in section 5. u_{pB} must be ranged between 0 and V; thus, the only solution of (38) is:

$$u_{pB} = \frac{1}{2Q}(-T + \sqrt{T^2 - 4QX}) \tag{39}$$

This solution is valid for any value of Q, different from zero.

If (A) and (B) are the same materials, we have $Q = 0$, and (38) has one solution which is:

$$u_{pB} = \frac{V}{2} \tag{39'}$$

The velocity and the pressure of the induced shock into (B) are given by:

$$U_{SB} = A_B + S_B u_{pB} \qquad P_B = \rho_{0B} U_S B u_{pB} \tag{40}$$

3)- pressure and velocities of the transmitted wave in medium (A) are given by:

$$\begin{aligned} P_A &= P_B = \rho_{0A} U_{SA} u_{pA} = \rho_{0B} U_{SB} u_{pB} \\ U_{SA} &= A_A + S_A u_{pA} \\ u_{pA} &= V - u_{pB} \end{aligned} \tag{41}$$

5)- coordinates of point N: Figure 19 shows that

$$x_n = -e + V t_n = -(U_{SA} - V) t_n$$

We get:

$$t_n = \frac{e}{U_{SA}} \quad \text{and} \quad x_n = e \left(\frac{V}{U_{SA}} - 1 \right) \tag{42}$$

6)- trajectory of wave NR.
We are in medium (A). The equation of the trajectory relatively to the frame at rest is:

$$\frac{x - x_n}{t - t_n} = C_A + u_{pA} \tag{43}$$

7)- coordinates of R.
At this point, the examination of Fig. 19 and equation (43) show that:

$$x_r = u_{pB} t_r \qquad \frac{x_r - x_n}{t_r - t_n} = C_A + u_{pA}$$

We obtain:

$$t_r = \frac{e}{U_{SA}} \frac{U_{SA} + C_A - u_{pB}}{C_A + u_{pA} - u_{pB}} \tag{44}$$

8)- at M, the release wave has overtaken the shock wave; the coordinates of M are given by:

$$x_m = U_{SB} t_m \qquad \frac{x_m - x_r}{t_m - t_r} = C_B + u_{pB} \tag{45}$$

t_m is the duration of the shock (pressure has not change between R and M); from this time, hydrodynamic decay begins and pressure decreases from P_B. With equations (44) and (45), we obtain:

$$t_m = \frac{e}{U_{SA}} \frac{C_A + U_{SA} - u_{pB}}{C_A + u_{pA} - u_{pB}} \frac{C_B}{C_B + u_{pB} - U_{SB}} \tag{46}$$

In the particular case where (A) is identical to (B), formula (46) gives:

$$t_m = \frac{e}{U_S} \frac{C + U_S - u_p}{C + u_p - U_S} \tag{46'}$$

(in equation (46') we have omit indexes A and B) (we can compute the trajectory of shock after time t_m [see Ribakov 1976]).

8.3 Numerical Examples

This example is taken from Melosh (1989).
The projectile is made of iron (we do not take into account the phase change of iron at 13 GPa.):
$A_A = 4.05$ km/s $\quad S_A = 1.41$
$\rho_{0A} = 7.8$ g/cm^3 $\quad \gamma_0 = 1.8$
The target is made of anorthosite (anorthosite is a mixture of silicates: albite ($NaAlSi_3O_8$) and anorthite ($CaAl_2Si_2O_8$) which are granites and basalts of continental crust of Earth):
$A_B = 7.71$ km/s $\quad S_B = 1.05$
$\rho_{0B} = 3.965$ g/cm^3 $\quad \gamma_0 = 1.1$
(this value is deduced from approximate law: $S = \frac{1}{2}(\gamma_0 + 1)$).
For an impact velocity of 5 km/s, (39), (40) and (41) give:
$u_{pB} = 2.83$ km/s $\quad u_{pA} = 2.17$ km/s
$U_{SB} = 10.7$ km/s $\quad U_{SA} = 7.1$ km/s
$P_B = P_A = 120$ GPa

For the evaluation of t_n, t_r and t_m, we want values of the sonic velocity at high pressure. For this, we use formula (36) and linear relationship between U_S and u_p; with the common approximation $\gamma/V = \gamma_0/V_0$, we obtain:

$$C = \frac{1}{\sqrt{A}}[A + (S-1)u_p]\sqrt{A + 2Su_p - \frac{\gamma_0 S u_p^2}{A + u_p}}$$

or:

$$C = \frac{1}{\sqrt{A}}[U_S - u_p]\sqrt{U_S + (U_S - A)(1 - \gamma_0 \frac{u_p}{U_S}} \tag{47}$$

For the considered impact, (47) gives:
$C_A = 10.1$ km/s $\quad$ and $\quad C_B = 7.1$ km/s
If projectile thickness is 1 km, we have:
$t_m = 1.11$ s $\quad$ and $\quad x_m = 11.8$ km
In the case of planar impact and only with the hydrodynamic decay, the shock propagates through 12 km without attenuation.

For this numerical application, we have taken a very low impact velocity: 5 km/s is valid only for the Moon !
For the meteorite of Rochechouard (France), they give (1993):
$V \approx 20$ km/s $\quad \rho_{0A} \approx 3.4$ g/cm^3 $\quad$ diameter ($\approx e$) ≈ 1.5 km
For the meteorite of Haughton (Canada) (Martinez 1993):
$V \approx 25$ km/s $\quad \rho_{0A} \approx 3.04$ g/cm^3 $\quad$ diameter ($\approx e$) ≈ 1 km
For an impact velocity of 30 km/s, our calculations give:
$P_A = P_B = 1920$ GPa $\quad$ and $\quad x_m = 4.2$ km

9 Equation of State and Temperature Evaluation

9.1 Mie-Gruneisen Equation of State

We have said (Sect. 4) that high pressures were associated with high temperatures but the knowledge of conservative equations and Hugoniot curve of a material is inadequate to compute temperature in a given shock. For this we must have an "equation of state": it is a relation between 3 thermodynamic variables P, V and E or T:

$$f(P, V, E \text{ or } T) = f(P_0, V_0, E_0 \text{ or } T_0) = 0$$

Numerous works have been devoted to this problem since the Second Word War. This problem is always important in geophycics and astrophysics.

The problem of equation of state and the applications to geophysics are enlarged in Zharkov & Kalinin (1971). Many developments in solid physics allow to compute equation of state of simple solids with a good accuracy. These calculations are founded on a good knowledge of electrical forces which act between charged ions on crystalline lattice sites and conduction electron gas. For more complex solids (this is always the case for geologic materials), equation of state is determined with experiments: the theory gives a general formula with unknow parameters which are computed to match with experimental data.

In the following, we want to study a widely used equation of state: the Mie-Gruneisen equation which is derived from the theory of specific heat from Debye. In this formulation the pressure and internal energy are separated into thermal (index T) and non-thermal (index K) parts:

$$P = P_T + P_K \tag{48}$$

$$E = E_T + E_K \tag{49}$$

$$P_K = -\frac{dE_K}{dV} \tag{50}$$

In these formula, P_K and E_K are functions of V only and P_T and E_T are functions of V and T.

The function $P_K(V)$ describes the isothermal compression at 0 K: it takes into account the part of ions and the part of electrons. Usually, one selects an empirical function which is more or less theoretically proved; this function has a lot of few parameters which are fitted with experimental data. If we know the Hugoniot curve, we will see that the knowledge of P_K (or U_K) is useless.

For moderate pressures (200 GPa in Copper, for example), we can consider only the thermal part of vibrating ions on the lattice sites. This part is given by the theory of Debye:

$$E_T = E - E_K = \int_0^T C_V dT \tag{51}$$

C_V is the specific heat at constant volume; E is the total internal energy of the solid. Thus, in this case, the equation of state of Mie-Gruneisen may be written as:

$$P = P_K + \frac{\gamma}{V}(E - E_K) \tag{52}$$

γ is the "Gruneisen constant" which depends only on volume. One can prove that γ is given by:

$$\gamma = \frac{\alpha K_T V}{C_V} \tag{53}$$

α is the coefficient of volumic dilatation and K_T the coefficient of isothermal rigidity:

$$K_T = -V(\frac{\partial P}{\partial V})_T \qquad \alpha = \frac{1}{V}(\frac{\partial V}{\partial T})_P \tag{54}$$

For the greater part of metals, γ is about 2 in the initial state. Usually, one takes the following approximation which is difficult to prove:

$$\gamma\rho = \gamma_0\rho_0 = cste \tag{55}$$

9.2 Temperature Behind the Shock Front. Low Pressure

In the range where this model of equation of state is valid, we can compute the temperature reached in a shock as follows. We start from the three following equations:

$$dE = TdS - PdV = C_V(dT + T\frac{\gamma}{V}dV) - PdV \tag{56}$$

$$P = P_K + \frac{\gamma}{V}(E - E_K) \tag{57}$$

$$E_H = E_0 + \frac{1}{2}(P_H + P_0)(V_0 - V) \tag{58}$$

Equation (56) is deduced from the second thermodynamic principle and the equation of state. Equation (58) is the conservative law of energy in a shock wave. Equation (57) is the equation of state valid for the shocked state (P_H,E_H,V). From the three Eqs. (56)-(58), we obtain the differential equation for temperature reached in the shock:

$$dT_H = \frac{V_0 - V}{2C_V}dP_H + [\frac{P_H - P_0}{2C_V} - \frac{\gamma}{V}T_H]dV \tag{59}$$

For this calculation, we assume that C_V is constant (T_H is greater than the Debye temperature).

For weak shocks, one can estimate temperature with the variation of entropy between initial and final states, ΔS_H. Equations (19) and (56) can be written:

$$\Delta S_H \approx \frac{1}{12T_0}(V_0 - V)^3(\frac{\partial^2 P_H}{\partial V^2})_0 \approx \frac{C_2S}{3T_0}(1 - \frac{V}{V_0})^3 \tag{60}$$

$$T_H \approx T_0 \exp[\frac{\Delta S_H}{C_V} + \gamma_0(1 - \frac{V}{V_0})] \tag{61}$$

C and S are the coefficients of the linear relationship between U_S and u_p.

9.3 Temperature Behind the Shock Front. High Pressure

In the range of high pressure, it is necessary to take into account of the energy absorbed by the conduction electrons. We must modify the equation of state in order to take into account this electronic gas; for this, we suppose that the coupling between electronic gas and ionic vibrations is negligible; we suppose also that the temperature reached in shock is not too high, so that the electronic gas is an ideal degenerate Fermi gas. From the first hypothesis we deduce that the thermal part of total internal energy is the sum of the energy of vibrating ions (51) and the kinetic energy of electronic gas; for the validity of second hypothesis, we must have:

$$T_H < T_{0F} = \frac{h^2}{8m_e k}(\frac{3N}{\pi V})^{2/3} \tag{62}$$

T_{0F} is called "the Fermi temperature". h is the Planck constant, k, the Boltzmann constant, m_e, the electronic mass and N, the ionic number in volume V.

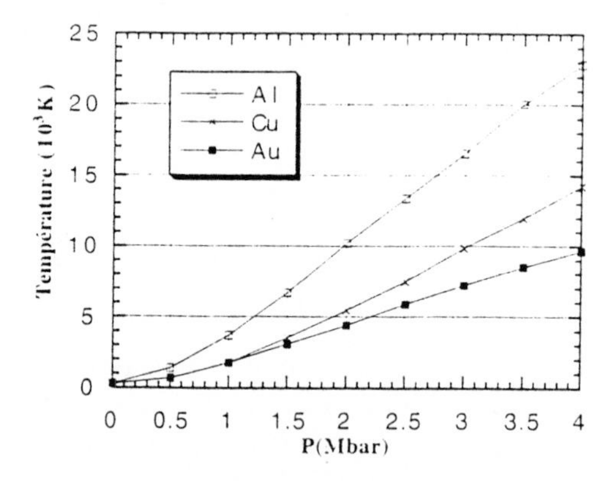

Fig. 21. Temperature vs. pressure in shock wave

For Copper, $T_{0F} \approx 7.1$ eV $\approx 82{,}000$ K.

In these conditions, the kinetic energy of electronic gas, E_{Te}, and its specific heat, C_{Ve}, are given by:

$$E_{Te} \approx \frac{1}{2}\beta_0(\frac{V}{V_0})^{2/3}T^2 \tag{63}$$

$$C_{Ve} \approx \beta_0(\frac{V}{V_0})^{2/3}T \tag{64}$$

$$\beta_0 = \frac{\pi^2(N/V_0)k}{2T_{0F}} \tag{65}$$

The electronic contribution in the thermal pressure is (for γ_e, Russian authors give 1/2. This value may be proven with the Thomas-Fermi-Dirac equation of state):

$$P_{Te} = \frac{\gamma_e}{V} E_{Te} \qquad \gamma_e = 2/3 \tag{66}$$

In these conditions, equation of state becomes:

$$P = P_K + \frac{\gamma}{V} E_{Tr} + \frac{\gamma_e}{V} E_{Te} \tag{67}$$

$$E = E_K + E_{Tr} + E_{Te} \tag{68}$$

E_{Tr} is the contribution of ionic vibrations given by equation (51). As before, we can obtain a differential equation for temperature calculations in a shock wave.

9.4 Numerical Applications

The two following figures show the results of temperature calculations in a shock wave. The first (Fig. 21) is in the case of high pressure with electronic contribution.

The Fig. 22 corresponds to the case of Aluminium. It shows the influence of electronic gas; at low pressure, this influence is negligible.

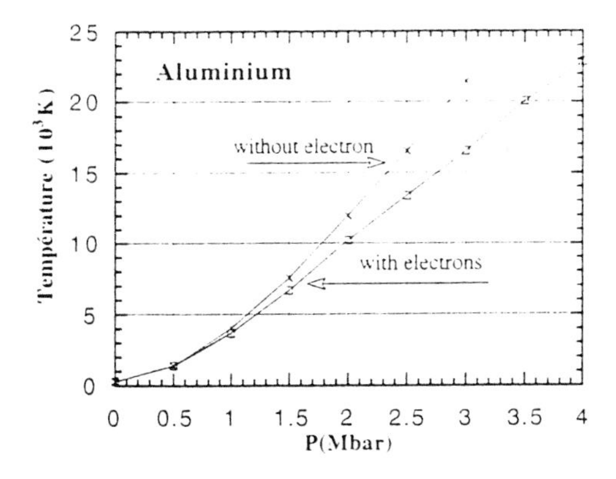

Fig. 22. Temperature vs. pressure: influence of electronic gas

9.5 Release from the Shocked State. Residual Temperature

In Sect. 6, we have studied the hydrodynamic decay of a shock and we have shown that the incident shock is weakened by sonic waves emitted at the end of the shock pulse,after a finite time which is a function of applied pressure. This damping is isentropic. It results that, after a shock, the solid expands from P_H (shock pressure) to P_0 (initial pressure before shock) and this expansion is isentropic, the entropy beeing S_H. S_H is the entropy of solid submitted to the pressure shock P_H. In (P,V) plane, the locus of succesive states taken by the solid during this expansion is above the Hugoniot curve (see Sect. 4): this transformation is a reversible one. The final state after expansion is P_0, $V_{0f} > V_0$, T_{0f}.

Now, we compute V_{0f} and T_{0f}. For this transformation, $dS = 0$ and (56) and (55) give:

$$\frac{dT}{T} = \gamma_0 \rho_0 dV$$

The integration between shocked state and release state gives:

$$T_{0f} = T_H \exp[-\gamma_0 \rho_0 (V_{0f} - V_H)] \tag{69}$$

If α is the volumic coefficient of dilatation at pressure P_0, we have:

$$V_{0f} - V_0 = \alpha V_0 (T_{0f} - T_0) \tag{70}$$

Equations (69) and (70) allow to compute T_{0f} and V_{0f}.

If C_P is the massive heat at constant pressure ($\neq C_V$), we can estimate the residual energy, E_{0f}, of final state with:

$$E_{0f} - E_0 \approx C_P (T_{0f} - T_0)$$

For example: for Aluminium shocked from the initial state $P_0 \approx 0$ GPa, $T_0 \approx 300$ K, (69) and (70) give:

P_H	V_H/V_0	$T_H - T_0$	$T_{0f} - T_0$
25 GPa	0.82	331 K	134 K
35 GPa	0.78	522 K	216 K

In Sect. 9.4, we have seen that the temperature reached in a shock wave may be high. Present calculations show that the residual temperature in an isentropic release may be important too.

9.6 Isentropic Curve from the Initial State

The knowledge of Hugoniot curve sprung from initial state (P_0,V_0,T_0) allows the computation of isentropic and isothermal curves sprung from this state if we know the general form of equation of state. We suppose that the material under study has a Mie-Gruneisen equation of state without electronic contribution.

For a shock from (P_0,V_0,T_0) to (P_H,V,T_H), (52) and (55) give:

$$P_H(V) = P_K(V) + \rho_0 \gamma_0 [E_H - E_K(V)] \tag{71}$$

For this same volume V, there is a state (P_S,V,T_S) on the isentropic curve sprung from the initial state. For that state, (52) and (55) give:

$$P_S(V) = P_K(V) + \rho_0 \gamma_0 [E_S - E_K(V)] \tag{72}$$

For the states H and S, we have:

$$P_S(V) = -\frac{dE_S}{dV} \qquad E_H = E_0 + \frac{1}{2})(P + P_0)(V_0 - V) \tag{73}$$

Equations (71)-(73) give:

$$P_H(V)[1 - \frac{\rho_0\gamma_0}{2}(V_0 - V)] = \rho_0\gamma_0 \int_{V_0}^{V} P_S(V)dV + P_S(V) \tag{74}$$

(74) is an integral equation which allows to compute $P_S(V)$ if we know $P_H(V)$. The knowledge of $P_K(V)$ is useless. Below, we give two analytical forms for $P_S(V)$ which are currently used.

1) The Murnaghan equation:

$$P_S(V) = \frac{K_0}{n}[(\frac{V_0}{V})^n - 1] \tag{75}$$

K_0 is the bulk modulus of isentropic rigidity in the initial state and n is a coefficient in relation with interatomic forces.

2)The Birch-Murnaghan equation:

$$P_S(V) = \frac{3}{2}K_0X^5(X^2 - 1)[1 + \frac{3}{4}(K'_{0P} - 4)(X^2 - 1)] \tag{76}$$

K'_{0P} is the derivative of K_0 with pressure and X is equal to $(V_0/V)^{1/3}$.

In these two equations, we have only two coefficients, K_0 and n (or K'_{0P}) which are computed with the Hugoniot curve P_H: in the initial state, P_H and P_S have the same first two derivatives with volume.

The Birch-Murnaghan may be deduced from the theory of finite deformations (Knopoff 1963).

9.7 Comments

In the region of very high pressure (2000 to 5 or 6000 GPa) we can use the Tillotson equation of state (Melosh 1989). This equation is a junction between the Mie-Gruneisen equation at low pressure and Thomas-Fermi-Dirac equation (Knopoff 1963) at ultra high pressure (above 10,000 GPa).

References

Boustie, M. (1991): *Etude de l'endommagement dynamique sous l'action d'une onde de choc induite par une impulsion laser de forte puissance dans une cible solide*, Thèse Université de Poitiers N°399 (11/01/1991).

Cottet, F. (1985): *Etude des caractéristiques d'une onde de choc de très haute pression induite par impulsion laser dans une cible solide*, Thèse Université de Poitiers N°407 (24/05/1985).

Doran, Linde (1966): Shock effects in solids. Sol. State Phys. **19**, 229.

Hayes, W.D. (1960): *Gas Dynamic Discontinuities*, Princeton University Press.

Knopoff, L. (1963): Equations of state of matter at ultra-high pressure. In Bradley R.S. (ed.) *High Pressure Physics and Chemistry*, Academic Press – London, New York.

Landau, L, Lifchitz, F. (1971): *Mécanique des fluides*, Editions MIR, Moscou.

Martinez, I. (1993): *Transformations de phases et libération de CO_2 lors d'un impact de météorite: étude des impactites du cratère de Houghton, Canada*, Thèse Université de Paris VII (26/11/93).
Melosh, H.J. (1989): *Impact Cratering. A Geologic Process*, Oxford Monographs on Geology and Geophysics N°11, Oxford University Press.
Ribakov (1976): Amortissement de l'onde de choc lors de choc entre plaques. Zh. Prikl. Mekh. Tekhn. Fiz. n° 5, p. 147 [in russian].
Smith, C.S. (1958): Metallographic Studies of Metals after Explosive Shock. Trans. Met. Soc. A.I.M.E. **214**, 574.
Zharkov, V.N., Kalinin, V.A. (1971): *Equations of state for solids at high pressures and temperatures*, Consultants Bureau - New York, London.
XXX (1993): Une météorite en France. Pour la Science (Juin), p. 23.

In the following, some references very useful in physics of shock waves.

Rice, M.H. et al. (1958): Compression of Solids by Strong Shock Waves. Sol. State Phys. **6**, 1.
Alt'shuler, L.V. (1965): Use of Shock Waves in High-Pressure Physics. Sov. Phys. Uspekhi **8**, 52.
McQueen, R.G. et al. (1977): The Equation of State of Solids from Shock Wave Studies. In Kinslow R. (ed.) *High Velocity Impact Phenomena*, Academic Press – London, New York.
Zel'dovich, Ya.B., Raiser, Yu.P. (1966): *Physics of Shock Waves and High-temperature Phenomena*, Academic Press – London, New York.
Murri, W.J. et al. (1974): Response of Solids to Shock Waves. In Wentorf R.H.Jr (ed.) *Advances in High Pressure Research*, Academic Press – London, New York.

The Experimental and Theoretical Basis for Studying Collisional Disruption in the Solar System

Donald R. Davis

Planetary Science Institute/SJI
620 North Sixth Avenue, Tucson, Arizona 85705-8331, USA

Bases expérimentales et théoriques pour l'étude des destructions collisionnelles dans le Système Solaire

Résumé. Les destructions collisionnelles catastrophiques ont joué un rôle majeur dans le façonnage de la population des astéroïdes et des objets de la ceinture de Kuiper au long de l'histoire du Système Solaire. Ce processus continue actuellement et est responsable des fragments envoyés vers le Système Solaire interne et la Terre, où ils se manifestent comme météores, comètes et NEAs. Ce chapitre décrit notre compréhension du processus de destruction collisionnelle résultant d'expériences d'impact au laboratoire, des modèles numériques d'extrapolation des résultats de ces expériences à des corps du même ordre de grandeur que les astéroïdes ainsi que des simulations d'évolution collisionnelle de populations de petits corps. S'y ajoutent des recettes numériques pour calculer le résultat de collisions énergiques entre corps de taille quelconque, qui puissent être appliquées à la modélisation des processus de fragmentation collisionnelle.

Abstract. Collisional catastrophic disruption has played a major role in shaping the population of asteroids and Kuiper belt objects over solar system history. This process continues to the present time and is responsible for launching fragments on their trip from the asteroid belt or the Kuiper belt to the inner solar system and Earth where they manifest themselves as meteors, comets and Near-Earth asteroids. This chapter describes our understanding of the collisional disruption process resulting from laboratory impact experiments, numerical models for extrapolating laboratory results to asteroid sized bodies and simulations of the collisional evolution of small body populations. Included are numerical recipes for calculating the outcome of energetic collisions between bodies of arbitrary size that can be used to model collisional fragmentation processes.

1 Introduction

What do meteors, whose beauty graces the sky on a clear, dark night, have in common with the cratered surface of the Moon? The answer is that both are the result of high-speed collisions between solar system bodies. Meteors show that this process continues to the present time, while the cratered surfaces of the Moon, Earth and all solid bodies of the solar system (except Io), attest to a long history of bombardment by high-speed impacts. The evidence for catastrophic disruption due to high-speed collisions is less direct, but there are abundant clues that the collisional disruption process is one that has played and continues to play a significant role in shaping our solar system. Asteroid families, statistically significant clumpings of asteroid orbit elements, are the observable fragments from the catastrophic disruption of large parent bodies that occurred over the past few billion years. The beautiful ring systems of the outer planets are thought to be the most spectacular manifestation of collisional disruption as small satellites were shattered and spread around a global band. The gossamer asteroidal dust bands discovered by IRAS provide evidence that collisional disruption fragments extend to dust particle sizes and that their production continues even to the present day.

Collisional disruption refers to the process by which energetic collisions break up a target body and disperse the fragments. Low energy impacts produce craters on the target; generally only a very small fraction of the volume of the target body is affected by the cratering event. Increasing the collisional energy shatters an increasing fraction of the body, causing large scale fracturing that can affect a significant fraction of the surface of the target and as much as perhaps 10% of its volume. We define these events to be sub-catastrophic impacts, based on a widely adopted definition of the threshold for catastrophic disruption, namely when the largest fragment from the collision contains 50% of the mass of the target. The collisional energy is a major factor in determining collisional outcomes; however, the way in which the energy is delivered also plays a role, as we shall see. We consider two regimes – hypervelocity impacts which occur when the impact speed exceeds the sound speed in the target material, and low velocity impacts when the collision speed is significantly less than the local sound speed.

Understanding the collisional evidence of the asteroids or the formation and evolution of ring systems requires a quantitative understanding of the process of collisional catastrophic disruption. The size of the bodies involved (up to hundreds of kilometers diameter) and the impact speeds (up to tens of km/s) preclude our being able to duplicate these events under controlled conditions. So, to develop a quantitative understanding of collisional disruption, we resort to carrying out high-speed impact studies in the laboratory, involving targets that are ≈ 10 cm in size and impact speeds up to ≈ 10 km/s. We then apply the data from these experiments to the study of solar system bodies through scaling laws, i.e., mathematical equations that specify how the process of catastrophic disruption scales with size of colliding bodies, their impact speed, the material of the bodies and its physical state, etc. The output from these scaling algorithms can then be compared with astronomical observations, both for the purpose of testing our methodology of studying catastrophic disruption and for learning about the evolution of our solar system. By the process of modeling different parts of the solar system we further our understanding of its formation and evolution.

In this chapter, I will provide an overview of laboratory impact experiments and the type of data that are gleaned from them. This will be followed by a summary of the methods used to develop scaling algorithms. Finally, I will outline a recipe for calculating the outcome of a catastrophically disruptive collision between any two bodies. This chapter concludes with a list of important areas for future work.

2 Laboratory Experiments

2.1 Experimental Facilities

The starting point for understanding catastrophic disruption by high-speed collisions is the laboratory: two-stage gas guns developed in the 1950's and 60's made possible experiments with impact speeds as high as 6 km/s. The Vertical Gun Range, developed by D.E. Gault in the early 1960's and located at NASA's Ames Research Center, was the first of these experimental facilities to be used for studying collisions involving solar system materials such as basalts. Other facilities, such as the gas gun of A. Fujiwara at Kyoto, Japan and that of T. Ahrens at the California Institute of Technology in Pasadena, CA, and the facility developed at Johnson Space Center by F. Hörz, have all contributed to developing a broad database of impact experiments. Many experimenters, whose results are given in the papers listed in the attached references, have used these impact facilities to develop the experimental database.

Two other types of laboratory facilities have contributed to the disruption database. First are explosive disruption experiments carried out in a travertine quarry near Montemerano (Tuscany), Italy, by G. Martelli and his colleagues (Martelli et al. 1993; Giblin et al. 1994). This facility is able to achieve much higher fragmentation energies, thus disrupting larger targets than can be achieved by high velocity impacts. Another advantage of the quarry is that experiments are done in an open environment, so there are no confining boundary conditions in the form of walls which can cause secondary fragmentation, thereby changing the original breakup fragmentation size distribution.

The other facility is the high pressure chamber constructed at the Boeing Company, developed by K. Housen and R. Schmidt (Housen et al. 1991). This unique chamber is able to be pressurized to about 40 MPa and can thereby simulate the effect of a confining overpressure on the fragmentation process, such as is found in the interior of large asteroids.

2.2 The Experiments

Collisional Outcomes. The outcome of a collision between two bodies depends on a variety of factors: The collision speed and impact angle, the material type and physical structure of the target and projectile, and the sizes of the bodies. There are a continuum of outcomes shown in Fig. 1, with one end member being inelastic rebound, i.e., both bodies survive the collision essentially intact. The other extreme is complete fragmentation, whereby both bodies are completely shattered with the fragments dispersed as individual entities.

From laboratory experiments we seek to answer many basic questions regarding fragmentation, such as:

- What are the ways in which bodies break up?
- How much collisional energy is needed to break up bodies of different composition and physical structure?
- How many fragments of different sizes are produced?
- What is the velocity distribution of the fragments?
- What is the distribution of fragment shapes?
- How fast do fragments spin?

Fragmentation Modes. There are four modes of fragmentation shown by experiments, depending on the speed of the collision and the target material (Fig. 2). High velocity impacts produce core type fragmentation in which the outer region of the target shatters into numerous fragments and spalls, leaving the central part of the target as a single core. On the other hand, at low impact speeds a cone type of fragmentation is observed in some, but not all, materials. In this mode, the volume around the impact point is shattered while the antipodal volume breaks up into several cone-shaped fragments with the point of the cone pointed in the direction of the impact point (see Fig. 2). The core and/or cone types of fragmentation have been observed for silicate, cement mortar targets and clay targets. A third type of fracture is observed for ice targets where the targets break into a few large pieces (Lange and Ahrens 1981; Kawakami et al. 1983). Finally, recent experiments into cooled iron meteorite targets (Ryan and Davis 1994) show an "excavation" type of fragmentation in which the excavated volume around the impact point increases with increasing energy, but the antipodal region is relatively undamaged. The reasons for the different types of fracture modes are not well understood but may be related to whether the impact speed is greater or lesser than the acoustic wave speed in the target material and the structure/geometry of the target.

Degree of Fragmentation. The degree to which a body is broken up by a collision depends on the energy of the collision, how this energy is partitioned among various forms (heating, fracturing, ejecta kinetic and rotational energy, etc.), and the material and physical structure of the target and projectile. A simple measure of the degree of fragmentation is provided by f_l, the ratio of the

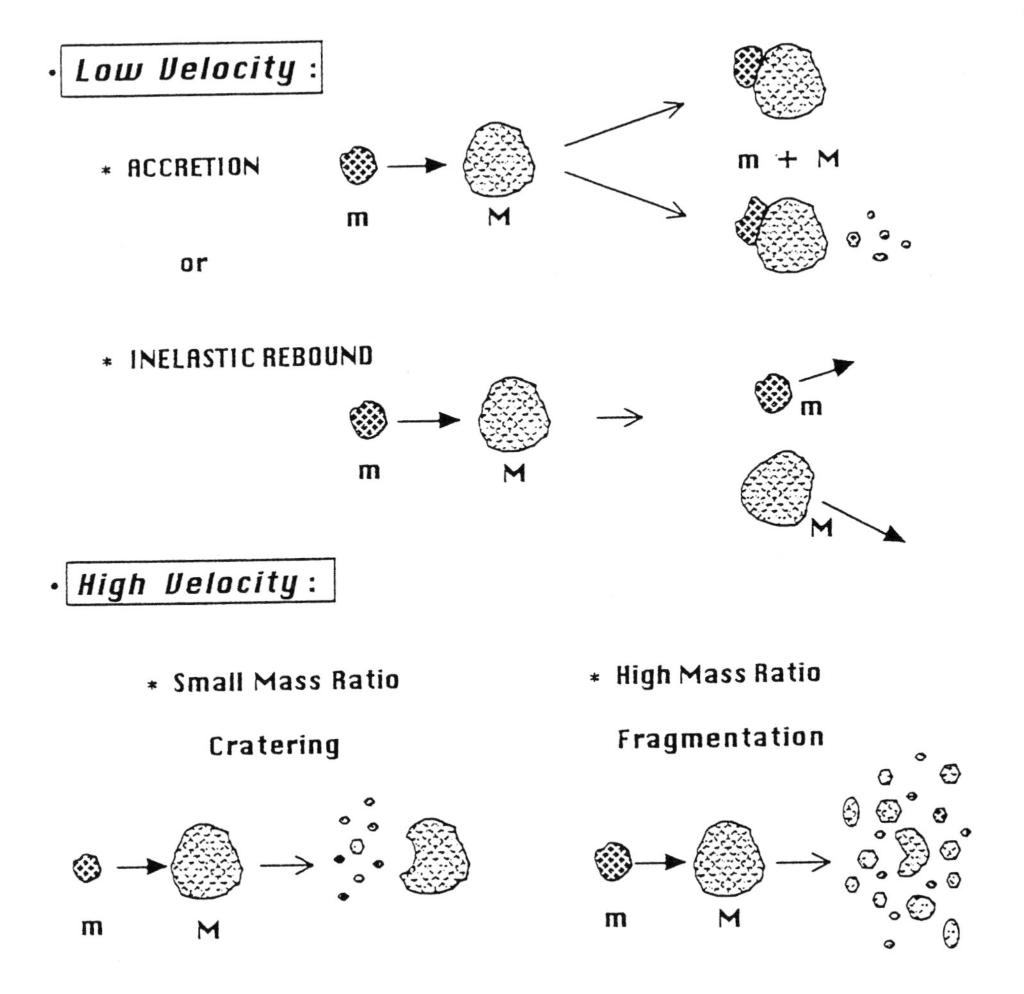

Fig. 1. Schematic illustration of the range of outcomes from collisions.

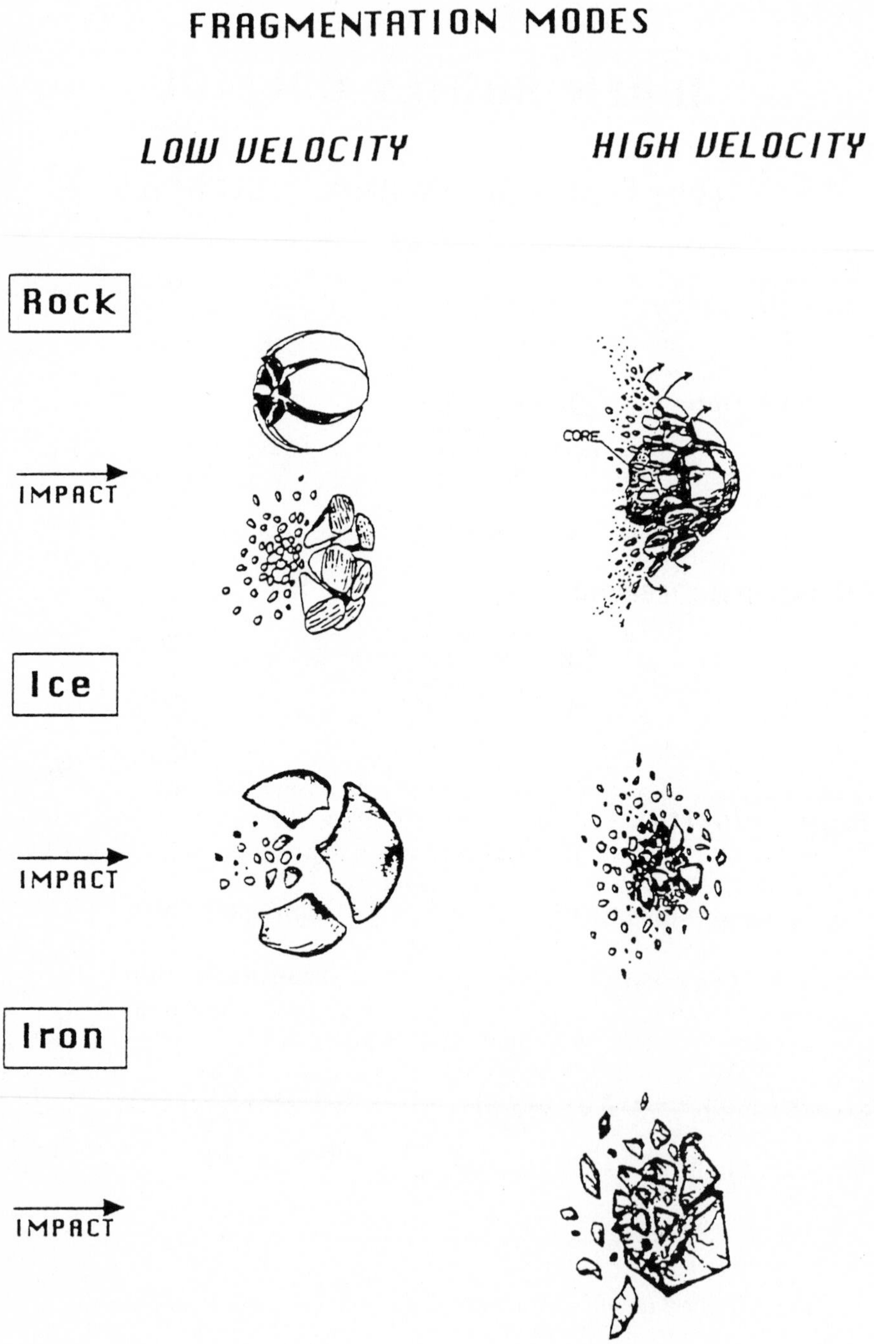

Fig. 2. Schematic illustration of the modes of collisional fragmentation.

mass of the largest fragment produced by the collision to the mass of the original body (M_L/M_T). By convention, a largest fragment containing 50% of the original mass defines the boundary between shattering and cratering collisions; collisions producing $f_l > 0.5$ are said to be cratering collisions, while those with $f_l < 0.5$ are shattering collisions.

The cratering regime as defined above includes much larger collisions than one normally regards as cratering impacts, and includes the transition impacts between normal cratering and shattering, that fracture between 10% and 50% of the mass of the target body. These transition impacts have not been well studied but are quite interesting in that they affect the global properties of a body, yet the target still retains the signature of the impact event, at least in some cases. The largest cratering events recognized in the solar system are the basin forming events on the Moon, where the projectiles were 100-200 km in diameter and the basins are over 1000 km in size, large craters on satellites of the outer planets, such as Mimas, Tethys, and Proteus, and the giant impact that formed the Vesta family of asteroids.

The shattering regime has been studied experimentally for a variety of target materials and impact speeds as shown in Fig. 3. Here the degree of fragmentation is shown as a function of Q, the specific collisional energy, i.e., the total collisional energy divided by the mass of the impacted body. Several facts are immediately apparent from this figure: One is the clear separation for the different types of material – there is relatively little overlap in Q needed to produce a given degree of shattering for different target material. Or, putting it another way, if you hit an unknown material with a known amount of energy, then you could identify with a high degree of certainty the type of target material from the mass of the largest fragment. We may quantify this observation using Q^*, the specific energy needed to produce a largest fragment containing 50% of the original target mass. Related to Q^* is the impact strength of the body, S, defined to be

$$S = \rho_T . Q^* \ , \tag{1}$$

where ρ_T is the density of the target body. Note that S has units of energy/volume, which is dimensionally equivalent to force/area, the customary units for measuring strength of material bodies. From Fig. 3, Q^*, is found to be 8.0×10^4 erg/g for ice, 3.2×10^6 erg/g for silicate, and about 6.8×10^8 erg/g for iron targets. Also seen from Fig. 3 is the correlation of f_l with Q – as expected, f_l decreases with increasing Q. Finally, one can see that there are other factors than just Q that determine f_l – note the range of values of f_l for a given material with the same value of Q.

The functional dependence of f_l on Q is determined experimentally by carrying out a series of impacts into the same type of target but varying Q in order to produce different degrees of fragmentation. Fujiwara et al. (1977) carried out just such a series of experiments and fit their data with an equation of the form

$$f_l = K \ Q^{-a} \ . \tag{2}$$

The parameter K varies from material to material and is a measure of the impact strength of the target while the exponent a is determined by the manner in which

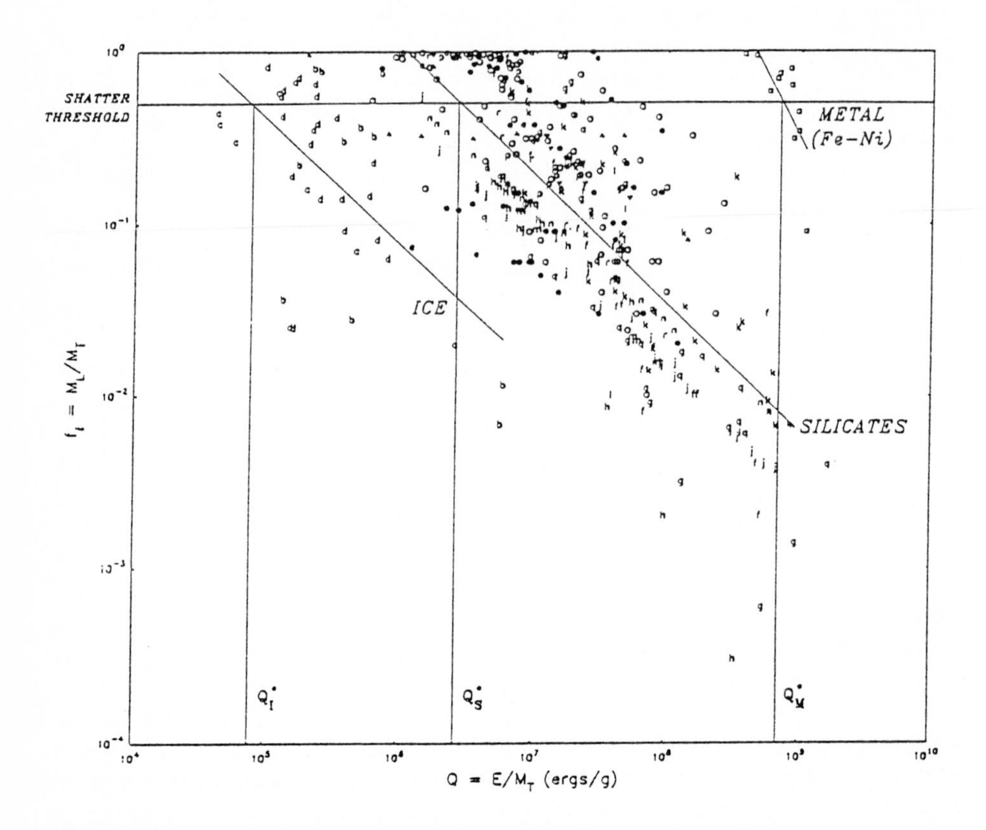

Fig. 3. Largest fragment mass (normalized to the mass of the original target), f_l, is shown as a function of the specific collisional energy (kinetic energy of the projectile divided by the mass of the target body) for three commonly found materials in the solar system. The value of Q that produces a largest fragment contains one-half of the mass of the original target is the shattering threshold (see text). References for the data points are: **Mortar**: a: Capaccioni et al. 1986; Davis & Ryan 1990; Ryan et al. 1991; Ryan 1992; ○: strong homogeneous; •: weak homogeneous; 6 : weak differentiated; 7 : pre-shattered; **Ice**: b: Lange & Ahrens 1981; c: Kawakami et al. 1983; d: Cintala et al. 1985; **Basalt**: e: Hartmann 1978; f: Fujiwara et al. 1977; g: Fujiwara & Tsukamoto 1980 (oblique); h, j: Matsui et al. 1982, 1984 (cube, sphere); i: Takagi et al. 1984; k: Takagi et al. 1984; l: Capaccioni et al. 1986; **Other materials**: m, n: Matsui et al. 1982 (granite, tuff); p: Fujiwara & Asada 1983 (clay); q: Matsui 1984 (tuff); r: Takagi et al. 1984 (pyrophillite); □: Ryan & Davis 1994 (iron). Additional information on the data shown in this figure can be found on the PSI home page on the World Wide Web (http://www.psi.edu).

the target fractures. For basalt, Fujiwara et al. (1977) found $K = 1.66\text{x}10^8$ erg/g and $a = 1.24$.

This empirical fit for the fragmental size distribution has been used by many workers in numerical simulations of collisional disruption (Greenberg et al. 1978; Davis et al. 1979). However, with the much larger database as shown in Fig. 3, fits can be made for three types of common solar system materials:

$$\text{Ice: } f_l = 1.7 \times 10^3 Q^{-0.72} \tag{3a}$$

$$\text{Silicates: } f_l = 6.0 \times 10^4 Q^{-0.78} \tag{3b}$$

$$\text{Iron: } f_l = 7.2 \times 10^{12} Q^{-1.49} \tag{3c}$$

Of interest is the similarity of the exponent for the ice and silicate fits ($a \approx 3/4$), substantially less than that for iron meteorites ($a \approx 3/2$). A caveat, though, is that there is much less data for the iron and the data covers a smaller range of f_l than is the case for silicates or ice, so the exponent could change as new experimental data become available.

Fragment Size Distribution. Numerous experiments show that the number of fragments increases rapidly with decreasing size and that a power law (or combination of power laws) gives a good representation of the fragment size distribution for all but the largest fragments (Hartmann, 1969; Fujiwara et al. 1989). Figure 4 shows the size distribution produced by several disruption experiments. There is a general tendency toward steepening of the size distribution with increasing degree of fragmentation as can be seen from Fig. 4a. A power law of the form

$$N(>m) = Cm^{-b} \tag{4}$$

is usually adopted to represent the cumulative size distribution of fragments.

FRAGMENT SIZE DISTRIBUTIONS

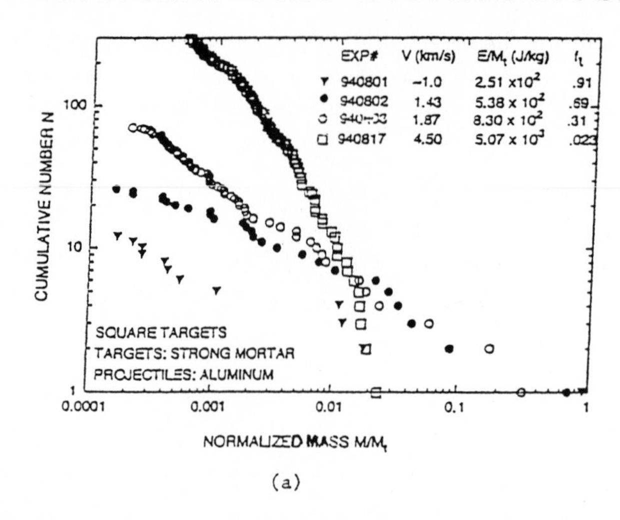

Fig. 4. Fragment size distributions produced by a wide range of collisional impact experiments, from large cratering impacts (a, 940801), to highly shattered targets (b, 820831). The trend toward steeper fragment size distributions (larger values of b with decreasing f_l) is shown in (a), while the small variation in the fragment size distributions from experiments having similar degrees of fragmentation is illustrated in (b). Experiments yielding a "two-slope" size distribution are shown in (c).

With this representation, the two unknowns of Eq. 4 (C and b) can be found from the mass of the largest fragment and the mass of the target. As shown by Greenberg et al. (1978):

$$b = 1/(1 + f_l) \tag{5}$$

and

$$C = m_l^b \tag{6}$$

These simple relationships have been widely used in studies of collisional fragmentation in the solar system.

Many experiments suggest that a single power law is an oversimplification for fragment size and that a two-slope model is a better representation of the experimental data. As shown in Fig. 4c, there is clear evidence in certain experiments for a change in slope in the fragmental distribution. Melosh et al. (1992) find a theoretical basis for this change of slope in the volume of material that is fractured at different distances from the impact site. Consider a disruptive impact into a finite, spherical target of radius R – the impact shock wave expands into the interior of the target and fractures an increasing volume of material with increasing distance from the impact site. However, when the shock front exceeds the distance $4R/3$ from the impact point, the volume of material fractured stops increasing with distance, and begins to decrease. It is this change in the volume of material that is fractured, which arises from the finite size of the target, that produces the two-segment size distribution of fragments.

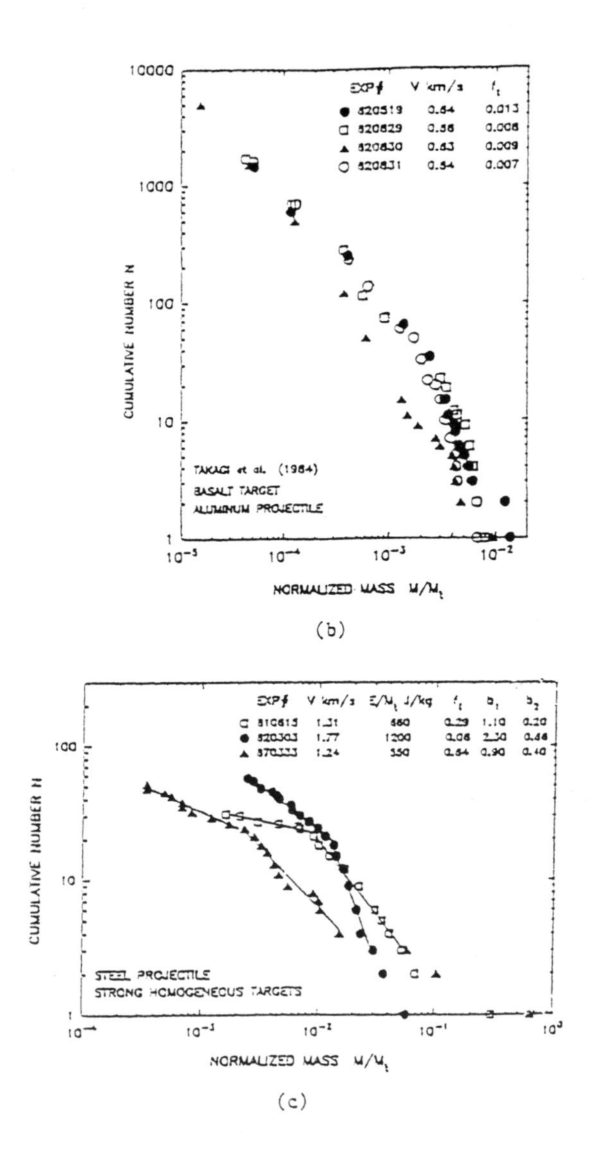

Fig. 4. (continued)

Fragment Shape Distribution. Fragments come in a wide range of shapes as shown in Fig. 5, yet there are certain general trends in fragment shapes. If a^* is the longest dimension of a fragment with b^* and c^* being the dimensions in the 2 directions orthogonal to a^*, then the dimensions $a^* : b^* : c^*$ are roughly in the ratio $2{:}\sqrt{2}{:}1$. No general trends in fragment shapes with fragment size have been found.

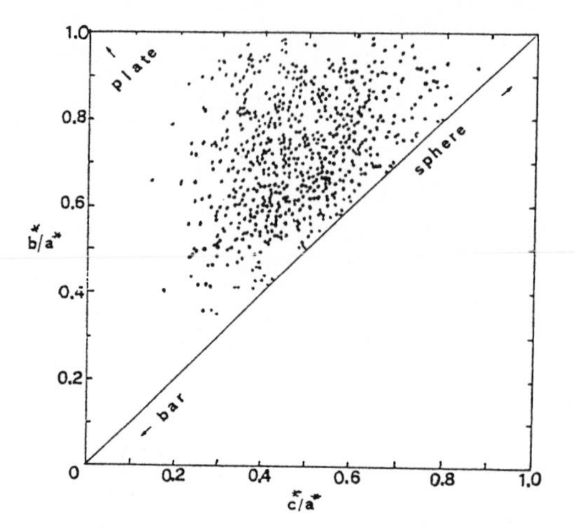

Fig. 5. Shape distribution of concrete fragments (after Capaccioni et al. 1984). a^*, b^* and c^* are the three orthogonal axes ($a^* \geq b^* \geq c^*$).

Fragment Spins. Fragments spin with a wide range of periods. The database on fragment spins from catastrophic breakup events is rather limited and deals primarily with basalt. The data suggests that there is an increase in spin rate with decreasing fragment size, although there is a wide range of rates for a given size fragment, as can be seen from the data of Fig. 6.

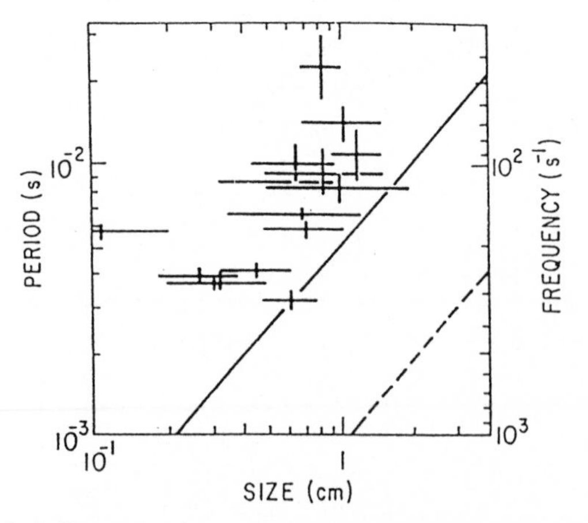

Fig. 6. Rotation period (frequency) as a function of fragment size. The horizontal bars represent the errors in the determination of fragment size and frequency. Solid line: Apparent minimum rotation period; broken line: rotational bursting limit for a sphere (figure from Fujiwara 1987).

Fragment Velocities. Fragment velocities are crucial for understanding many solar system phenomena and data from laboratory experiments is critical. Yet the data are relatively sparse, due in no small part to the difficulty in measuring the velocity of a particular fragment from high-speed films taken during the breakup event. Two orthogonal views are needed to find the velocity vector of a fragment, yet it is frequently difficult to recognize a fragment from different perspectives, partly due to the optically thick dust/small fragment cloud that accompanies the disruption. The largest fragments can usually be identified, yet the 20-30% of the target mass that is in the smallest fragments is generally traveling the fastest and is the most difficult to measure.

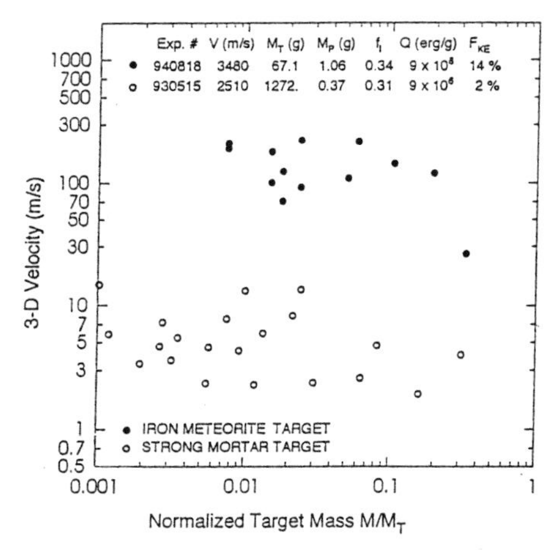

Fig. 7. Mass-velocity distribution of fragments from impact experiments having (a) large Q (940818), (b) moderate Q (930515). The degree of fragmentation was similar in the two experiments ($f_l = 0.34$ and 0.31, respectively), but the specific energy needed to shatter the iron target is 100 times that needed for silicate, leading to substantially higher fragment velocities for the iron fragments.

This point is one of the general observations about fragment speeds – typically the largest fragments are moving slowest, although there is considerable scatter about this general trend (Nakamura et al. 1992). Another general trend is that fragment velocities increase with Q, the specific energy of the collision. Figure 7 illustrates this trend: A high Q impact into iron produces fragment speeds of hundreds of m/s, while a more modest Q impact into silicates generates fragments moving at tens of m/s.

Fragment Mass-Velocity Relationship. Ideally, we would like to measure the mass and velocity of every fragment from a catastrophic breakup event, but as noted above this is currently impossible. In practice, we are forced to estimate the mass-velocity ($m - V$) distribution from a sample of the fragments. The earliest determination of the $m - V$ relationship, and one that has been widely adopted over the past three decades, was that by Gault, Shoemaker and Moore (1963), who presented data on the mass-velocity distribution from cratering impacts. The data can be modeled as:

$$f(>V) = (V/V_{min})^{-k} \text{for} V \geq V_{min} \tag{7}$$

where $f(>V)$ is the fraction of ejecta mass moving faster than speed V, V_{min} is a parameter which depends on the energy of the impact, and k ($\approx 9/4$) is the slope of the $m-V$ distribution. This distribution, although measured only for ejecta from cratering impacts, has been widely adopted to describe the $m-V$ distribution from shattering impacts as well. One difficulty with Eq. 7 is that it gives the mass fraction moving faster than a given speed, not the velocities of individual fragments that are required for studying the formation of asteroid families or secondary crater formation on planetary surfaces. Recent experiments by Nakamura and Fujiwara (1991) found a degree of correlation between the mass and velocity of fragments which they expressed as:

$$V(m) = V_0 \left(\frac{m}{M}\right)^{-r} , \tag{8}$$

where M is the target mass and V_0 is a parameter determined by the specific energy of the experiment (Q). The exponent, r, is related to the exponent k of Eq. 7 by $r = (1-b)/k$, where b is the slope of the fragment size distribution as given by Eq. 5. Marzari et al. (1995) applied Eq. 8, with a suitable random variation in the velocity, to model the formation and subsequent evolution of asteroid families.

Energy-Partitioning for Fragment Speeds. By integrating the $m-V$ distribution over all fragment masses, one can find the total kinetic energy carried by these fragments, and thus determine f_{KE}, the fraction of the collisional kinetic energy that is partitioned into fragment translational motion. Ejecta kinetic energy is only one of the modes that the collisional energy goes into; others are comminution, heat and rotational motion. In high velocity impacts, a few percent of the collisional energy goes into ejecta kinetic energy, while a larger fraction of the collisional energy is so partitioned in low velocity collisions, perhaps as high as 10-20%. One caveat to be aware of is how f_{KE} is measured: Some workers use the total kinetic energy in the collision, while others assume that the collisional energy is divided between the target and projectile. How the energy is divided is a critical question, and it is frequently assumed that it is split evenly between target and projectile. This assumption has recently been shown to be valid when the target and projectile are made of the same material (Hartmann 1980, 1988; Ryan and Davis 1994), but is probably not true for dissimilar materials. Much more work is needed to understand the fragmentation process when there is a material difference between the target and projectile.

3 Scaling Algorithms and Fragmentation Outcome Models

We now turn to the problem of how to apply the laboratory results presented above to problems involving real solar system bodies. The major difficulty is one of scale: Laboratory experiments involve targets that are several tens of cm in size at largest, while asteroids and satellites range up to 1000 km in diameter, a factor of 10^6-10^7 bigger. Given that we know the energy needed to break up a 10-cm target, how do we calculate the energy required to shatter a body that is a million times larger?

The first solution to this problem was to assume that the specific energy needed for fragmentation, Q^*, was independent of the size of the body, the so-called energy scaling. Thus Q^* needed to break up a silicate asteroid was the same as the Q^* needed to break up a target having the same composition in the laboratory. By disrupting a variety of silicate targets (and simulants of silicates), one can experimentally determine the energies needed to fragment asteroids of different compositions.

Energy scaling was found to be inadequate when applied to calculating the size distribution of asteroid families by Davis et al. (1985). The energy needed to disperse the fragments produced by breakup of large asteroid parent bodies far exceeds the energy needed to shatter the asteroid when energy scaling is assumed. Thus models assuming energy scaling generally predicted sizes of asteroid family members that were much smaller than actually observed, clearly indicating that something was wrong with the scaling. The solution proposed by Davis et al. (1985) was that of gravitational self-compression, whereby the energy needed to shatter a body for which self-gravity is significant increases with the size of the body. This is due to the gravitational compression produced on the interior of a body from the weight of the overburden material. Since fracture generally occurs due to failure in tension, then the tensional stress must exceed the sum of the intrinsic tensile strength of the material plus the compressive stress produced by the overburden. The effective Q^* as a function of body size is shown in Fig. 8 and such a model predicts asteroid family sizes that are in much better agreement with the observed sizes than does the energy scaling model without the self-compression effect.

The above models are empirical ones. In an attempt to calculate Q^* based on physical principles, Farinella et al. (1982) proposed that the fracture energy should be proportional to the total surface area of the fragments, an idea based on the concept that there are a constant number of mechanical bonds across a unit area of material, so the greater the surface area of the fragments, the bigger the number of bonds that have to be broken. Using this model, Farinella et al. found that Q^* should vary as $1/\sqrt{D}$, where D is the diameter of the target body. Thus big asteroids (or satellites) are relatively easier to shatter than are small ones. This model, though, did not include gravitational compression, so the decrease continued to indefinitely large bodies.

Concepts from modern fracture theory which had been used in cratering studies were first applied to the problem of collisional shattering by Holsapple and Housen (1986). The key elements of fracture theory are summarized here; for a more detailed exposition, see Melosh (1989):

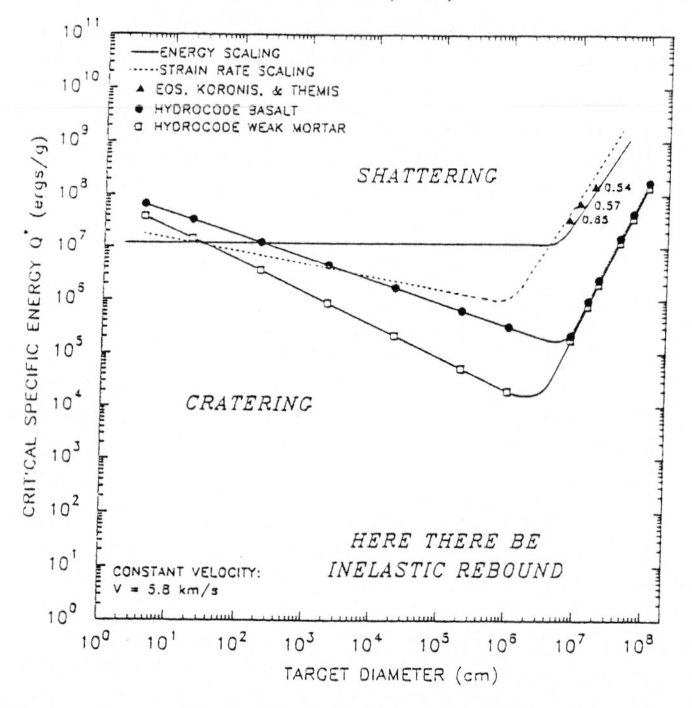

Fig. 8. Scaling laws for the variation of the critical specific energy needed to fracture silicate targets (see Table 1). The value of f_l for the Eos, Koronis and Themis Hirayama families is shown adjacent to their data points. Figure adapted from Davis et al. (1994) and Housen and Holsapple (1990).

* All natural materials are flawed, i.e., they contain a large number of cracks of differing sizes. The size distribution of cracks can be described by an power-law distribution:

$$N = L^{m/2} , \tag{9}$$

where L is the length of the crack and m is an experimentally determined parameter.

* When a material is subjected to a tensile stress, these pre-existing flaws begin to increase in length, i.e., to grow. There is a threshold stress, σ_L, which causes a crack to initiate growth given by:

$$\sigma_L = 1/\sqrt{L} \tag{10}$$

Hence larger cracks initiate growth at lower stress levels than do smaller flaws.

* The number of flaws that are activated at a given stress level is given by the Weibull distribution:

$$N = k\ \epsilon^m\ , \tag{11}$$

where ϵ is the stress.
* A crack is able to relieve the stress in a volume whose characteristic size is the length of the crack. Thus as a crack grows, it relieves the stress over an increasing volume of the material.
* The speed with which a crack lengthens is a fraction of the sound speed, c, in the medium, i.e.,

$$V_c = K_c \cdot c\ . \tag{12}$$

Typically, $K_c = 0.4$.
* The tensional stress that a volume experiences due to a high-speed collision is time dependent with an initial steep rise to a maximum value which depends on position within the body, followed by a decrease over a long time.

A high-speed collision produces both compressive and tensional stress waves in the interior of bodies, the latter being produced by reflections of compressive waves at free surfaces. Tensional failure is produced when the tensional stress exceeds the material strength (plus any compressional stress) and causes cracks to propagate throughout the volume that is under stress. The higher the stress level, the smaller the crack that can grow and since there are many more small cracks than large ones, are many more cracks activated at high stress levels than at low ones. Cracks continue to expand as long as the stress exceeds the threshold for crack activation. Thus large cracks grow for longer times than do small ones. If the stress remains high so that cracks can coalesce over the volume of the body, then the body experiences impact fracture and the minimum impact energy that causes this condition defines Q^*. Impact energies greater than this minimum value activate even smaller cracks, thus allowing these flaws to coalesce over smaller volumes, resulting in a greater degree of fragmentation for the target. This is why, for example, f_l defined above decreases with increasing Q, as empirically described by Eq. 2.

In the strain-rate scaling theory of Housen and Holsapple (1986, 1990), the concepts of fracture theory are coupled with the methodology of dimensional analysis to develop scaling equations for Q^*. Two physical regimes are treated. First is the strength regime, where the material properties of the bodies are dominant. The other is the gravity regime where the compressive stresses due to gravitational overburden dominate over material strength. Their strain-rate scaling equations are given in Table 1, and Fig. 8 shows Q^* as a function of target size assuming a basaltic composition. For this material, the transition from strength scaling to gravity scaling occurs for bodies 10-20 km diameter. Note also that for a fixed impact speed, Q^* decreases with increasing target size as $D^{-0.24}$, nearly as the fourth root of the size. Big bodies are relatively easier to break up than small ones and there is over an order of magnitude decrease in Q^* as we go from 10 cm laboratory sized targets to 10 km diameter asteroids.

Table 1. Scaling Equations for Impact Strength. The following equations, taken from Davis et al. (1994), give the impact strength as a function of target radius R for impacts at the mean asteroidal impact speed of 5.8 km/s. Cgs units are used. S_0 is the impact strength for 10 cm targets and K is the self-compression constant (see Davis et al. 1994).

Energy scaling:

$$S = S_0 + \frac{4\pi K G \rho^2 R^2}{15} .$$

Strain rate scaling:

$$S = S_0 \left(\frac{R}{10\ \mathrm{cm}}\right)^{-0.24} \left[1 + 2.14 \times 10^{-11} K \left(\frac{R}{10\ \mathrm{cm}}\right)^{1.89}\right] .$$

Hydrocode-basalt:

$$S = S_0 \left(\frac{R}{10\ \mathrm{cm}}\right)^{-0.43} \left[1 + 1.07 \times 10^{-17} \left(\frac{R}{10\ \mathrm{cm}}\right)^{3.07}\right] .$$

Hydrocode-weak mortar:

$$S = S_0 \left(\frac{R}{10\ \mathrm{cm}}\right)^{-0.61} \left[1 + 2.00 \times 10^{-17} \left(\frac{R}{10\ \mathrm{cm}}\right)^{3.25}\right] .$$

More detailed calculations of collisional disruption can be achieved by using numerical hydrocodes, suitably modified to include material strength effects. Numerical hydrocodes have been developed which use continuum mechanics and equations of state to calculate the response of a material, assumed to behave as a fluid, to a stress wave. Melosh et al. (1992) pioneered the inclusion of material strength and an explicit fracture model based on modern fracture theory into a two-dimensional hydrocode. Using this tool, they were able to reproduce most features of laboratory collisional disruption experiments.

Another approach to modeling collisional disruption uses the methodology of smooth particle hydrodynamics. The target is approximated by a large number of "particles", each of which represents a small volume of the target body. Each particle carries all of the physical variables – mass, position, velocity, density, etc. – needed to characterize the hydrodynamic state of the body. Benz and Asphaug (1994) added variables representing finite strength effects for solids to the "particles," thus creating a code for studying collisional disruption. This approach is currently undergoing validation against laboratory experiments, but offers great promise, particularly when coupled with increasingly powerful computers, of providing detailed simulations of collisional disruption.

Finally, a semi-empirical model for disruption has been developed by Paolicchi et al. (1989) and Verlicchi et al. (1994), who assumed that an impact produces a position dependent velocity field throughout the volume of the target. Stresses within the target body are generated by the gradient of this velocity field. Fracture occurs when these stresses exceed the specified criterion for strength of the target. The velocity gradient also accelerates the fragment and causes it to rotate; both the ejecta velocity and spin rates of fragments are calculated in this model. Empirical parameters of the model are found based on laboratory experiments.

4 Numerical Algorithms for Calculating Collisional Outcomes

Given that a collision occurs between two bodies at a given speed, how do we calculate the outcome of that collision? Such algorithms are used in studies of asteroid collisional evolution, the formation of asteroid families, the breakup of satellites to form ring systems or to be reaccreted into a "born-again" satellite. We now outline a numerical recipe for calculating collisional outcomes.

Such a semi-empirical model is much faster to calculate than are hydrocode simulations, thus allowing exploration of "impact parameter" space in a reasonable amount of time. This approach gives a tool intermediate between analytic scaling laws and numerical hydrocode calculations with which to study collisional disruption on many different size scales.

The first step is to classify the type of collision. To do this, we need to know the type of material that the target is made of, and its physical state, e.g., whether it is a competent body or has a fractured, rubble-pile structure. These quantities, when coupled with a scaling law, enable the impact strength of arbitrary-sized bodies to be calculated. A modified form of Eq. 3 that replaces the numerical constant with impact strength is:

$$f_l = 0.5 \ (S/(\rho * Q))^{-a} \tag{13}$$

This equation for f_l is based on central impacts, whereas most impacts in nature occur at oblique angles. A more refined calculation (Davis et al. 1985), computes an effective value of f_l, f_l^*, by averaging over all impact angles, giving:

$$f_l^* = 3 \ f_l^{2/3} - 2f_l \tag{14}$$

Thus an oblique impact produces a bigger largest fragment than does a central impact at the same specific energy.

Knowing f_l^*, we can now classify the collision: For $f_l^* \leq -0.5$, it is a shattering one; otherwise it is non-shattering, or a cratering impact. For shattering collisions, we calculate the size distribution for the fragments from Eqs. 4, 5 and 6.

The above algorithm assumes that fragments are distributed over all sizes down to zero. In the real world, of course, there is some lower bound on the size distribution. The resulting truncated mass distribution, properly referred to as a Pareto distribution (Cellino et al. 1991), provides a more refined representation of the fragment size distribution, although with the penalty of additional mathematical complexity (Petit and Farinella 1993).

As noted above, many impact experiments produce a fragment distribution that is better represented by a 2-segment power law than by Eq. 4. Again, we are faced with a more realistic model at the expense of simplicity. The single power-law model has two parameters – C and b – which are easily found as described earlier. The two-segment fit adds two more parameters, namely a second slope parameter and the mass at which the slope change occurs. Without a good theoretical or experimental basis for finding the additional quantities, they essentially become free parameters of the model. Until a better understanding of the two- slope model is achieved, the single slope model is preferred for most modeling applications.

Knowing the fragment sizes, we next need to calculate how fast these fragments are moving. Using Eq. 7 with $V = V_{esc}$, the effective escape speed of the target, we can find the amount of ejecta that escapes from the gravity field of the target, e.g.,

$$f_{esc} = (V_{esc}/V_{min})^{-k} \quad \text{if} \quad V_{esc} > V_{min}, \tag{15}$$

otherwise $f_{esc} = 1$.

However, this approach does not specify what are the sizes of escaping fragments. To find fragment sizes explicitly, we assume that smaller fragments travel the fastest. This assumption is borne out experimentally, although the correlation is not perfect (Petit and Farinella 1993). So if f_{esc} of the disrupted body escapes, the sizes are found by integrating the size distribution from the smallest fragments up to the largest escaping fragment, m_k, whose mass is given by:

$$m_k = m_l \left[\frac{f_{esc}}{f_l^*}(1-b)/b \right]^{1/(1-b)} \quad . \tag{16}$$

The largest escaping fragment, which barely escapes, has zero relative speed far from the reaccumulated body. Smaller fragments have increasingly large relative speeds, given by:

$$V = (V^2(m) - V_{esc}^2)^{1/2} , \tag{17}$$

where $V(m)$ is given by Eq. 8. Assuming a radial velocity field and by randomly assigning directions to the velocities, one can then calculate the orbit of the fragment, given the position in the orbit of the target body where the collision occurs (see Marzari et al. 1995 for a detailed description of this procedure).

5 Areas for Future Work

When we apply the above algorithm to solar system problems, discrepancies appear. First is the "velocity dilemma." Orbital elements for asteroid families imply that separation velocities of fragments from catastrophic disruption are typically hundreds of meters/sec. Laboratory measurements show fragment speeds of tens of meters/sec, about an order of magnitude smaller than what is inferred for the asteroid families. Perhaps there is some scaling of fragment speeds with body size so that pieces of disrupted asteroids move faster than do pieces of laboratory targets. But further work is needed to elucidate the origin of this discrepancy. Another area for work is to extend the experimental database. Of course it is a standard cry of the scientist that just a bit more data will shed light on an array of dark mysteries. Catastrophic disruption studies are no exception; however, there is a wide range of impact conditions in the solar system and many important arenas remain to be explored.

One important area is to use realistic solar system materials, i.e., meteorites and their simulants, for impact experiments. Initial work in this area has been done by Matsui and Schultz (1984) and Ryan and Davis (1994), using iron meteorite targets and by Cintala et al. (1995) for ordinary chondrite meteorites. However, many asteroidal materials remain to be studied to determine impact strengths and fragment speeds.

Related to using realistic materials is to use realistic shapes, particularly for small asteroids. Most impact experiments used spherical targets, which are good approximations for larger asteroids, but poor ones for small ones, as the Galileo images of Gaspra and Ida have shown. So, studying the effect of irregular-shaped targets on collisional outcomes is another area for future study.

A large challenge for future experimenters will be to validate the scaling algorithm. As discussed in Section 3, several scaling laws have been put forth describing how impact strength varies with target size. Discriminating among these is not an easy task, but disrupting large targets is one way of testing scaling laws. Doing this, however, is not easy – it requires the fragmentation of targets as large as a few meters. The energy needed is about three orders of magnitude larger than what is achievable from current experimental guns. So, either much larger guns are needed or another technique, such as explosive disruption, is required to break up such large targets. In any event, this will present a serious challenge for future researchers.

A final topic for further study is how fragments spin, particularly from disruption of bodies of different materials and sizes. Initial work on fragment spins was carried out by Fujiwara and Tsukamoto (1981) and more recently by Nakamura et al. (1992). However, studies of the collisional evolution of asteroid spins by Davis et al. (1989) are not in agreement with the observed spins, strongly suggesting that the modeling of asteroid spins is incompletely understood. Understanding the spin rates for a collisionally evolved population remains a challenge.

While we have dealt with the catastrophic disruption of laboratory scale targets and solar system bodies, the possible applications of fragmentation theory extend far beyond this limited domain. A grand application of fragmentation theory is given by Brown et al. (1983), who ask if the size distribution of galaxies is consistent with their formation by a fragmentation event, a.k.a. the Big Bang. They found that a power-law represents quite well the number of galaxies larger than a given mass. While this is not really evidence for a "Big Bang" origin of the universe, it is interesting to speculate on the processes that led to such a mass distribution for galaxies. But I shall leave that topic to the next symposium.

Acknowledgements. Eileen Ryan provided stimulating discussions, valuable criticisms and careful assistance in preparing figures, all of which contributed significantly to improving this paper. I also thank Paolo Farinella and Francesco Marzari for helpful comments on the manuscript. This paper was supported under NASA grant NAGW-4313.

References

Benz, W., Asphaug, E. (1994): Impact simulations with fracture. I. Methods and tests. Icarus **107**, 98–116.

Brown, W.K., Karpp, R.R., Grady, D.E. (1983): Fragmentation of the universe. Astrophys. and Space Sci. **94**, 401–412.

Capaccioni, F., Cerroni, P., Coradini, M., Farinella, P., Flamini, E., Martelli, G., Paolicchi, P., Smith, P.N., Zappalà, V. (1984): Shapes of asteroids compared with fragments from hypervelocity impact experiments. Nature **308**, 832–834.

Capaccioni, F., Cerroni, P., Coradini, M., Di Martino, M., Farinella, P., Flamini, E., Martelli, G., Paolicchi, P., Smith, P.N., Woodward, A., Zappalà, V. (1986): Asteroidal catastrophic collisions simulated by hypervelocity impact experiments. Icarus **66**, 487–514.

Cellino, A., Zappalà, V., Farinella, P. (1991): The size distribution of main-belt asteroids from IRAS data. Mon. Not. R. Astr. Soc. **253**, 561–574.

Cintala, M.J., Hörz, F., Morrison, R.V., See, T.H., Vilas, F. (1995): The response of chondritic targets to impact. Lunar Planet. Sci. Conf. **26**, 247–248.

Cintala, M.J., Hörz, F., Smrekar, S., Cardenas, F. (1985): Impact experiments in H_2O ice. II: Collisional disruption. Lunar Planet. Sci. Conf. **16**, 1298–1300.

Davis, D.R., Chapman, C.R., Greenberg, R., Weidenschilling, S.J., Harris, A.W. (1979): Collisional evolution of asteroids: populations, rotations, and velocities. In Gehrels T. (ed.) *Asteroids*, Univ. of Arizona Press, pp. 528–537.

Davis, D.R., Chapman, C.R., Weidenschilling, S.J., Greenberg, R. (1985): Collisional history of asteroids: Evidence from Vesta and the Hirayama families. Icarus **62**, 30–53.

Davis, D.R., Ryan, E.V. (1990): On collisional disruption: Experimental results and scaling laws. Icarus **83**, 156–182.

Davis, D.R., Ryan, E.V., Farinella, P. (1994): Asteroid collisional evolution: Results from current scaling algorithms. Planet. Space Sci. **42**, 599–610.

Davis, D.R., Weidenschilling, S.J., Farinella, P., Paolicchi, P., Binzel, R.P. (1989): Asteroid collisional history: Effects on sizes and spins. In Binzel R.P., Gehrels T., Matthews M.S. (eds.) *Asteroids II*, Univ. of Arizona Press, pp. 805–826.

Farinella, P., Paolicchi, P., Zappalà, V. (1982): The asteroids as outcomes of catastrophic collisions. Icarus **52**, 409–433.

Fujiwara, A. (1987): Energy partition into translational and rotational motion of fragments in catastrophic disruption by impact: An experiment and asteroid cases. Icarus **70**, 536–545.

Fujiwara, A., Asada, N. (1983): Impact fracture patterns on Phobos ellipsoids. Icarus **56**, 590–602.

Fujiwara, A., Cerroni, P., Davis, D.R., Ryan, E.V., Di Martino, M., Holsapple, K., Housen, K. (1989): Experiments and scaling laws for catastrophic collision. In Binzel R.P., Gehrels T., Matthews M.S. (eds.) *Asteroids II*, Univ. of Arizona Press, pp. 240–265.

Fujiwara, A., Kamimoto, G., Tsukamoto, A. (1977): Destruction of basaltic bodies by high-velocity impact. Icarus **31**, 277–288.

Fujiwara, A., Tsukamoto, A. (1980): Experimental study on the velocity of fragments in collisional breakup. Icarus **44**, 142–153.

Fujiwara, A., Tsukamoto, A. (1981): Rotation of fragments in catastrophic impact. Icarus **48**, 329–334.

Gault, D.E., Shoemaker, E.M., Moore, H.J. (1963): Spray ejected from the lunar surface by meteoroid impact. NASA Tech. Note D-1767.

Giblin, I., Martelli, G., Smith, P.N., Cellino, A., Di Martino, M., Zappalà, V., Paolicchi, P. (1994): Field fragmentation of macroscopic targets simulating asteroidal catastrophic collisions. Icarus **110**, 203–224.

Greenberg, R., Wacker, J.F., Hartmann, W.K., Chapman, C.R. (1978): Planetesimals to planets: Numerical simulation of collisional evolution. Icarus **35**, 1–26.

Hartmann, W.K. (1969): Terrestrial, lunar and interplanetary rock fragmentation. Icarus **10**, 201–213.

Hartmann, W.K. (1978): Planet formation: Mechanism of early growth. Icarus **33**, 50–61.

Hartmann, W.K. (1980): Continued low-velocity impact experiments at Ames Vertical Gun Facility: Miscellaneous results. Lunar Planet. Sci. Conf. **11**, 404–406.

Hartmann, W.K. (1988): Impact strengths and energy partitioning in impacts into finite solid targets. Lunar Planet. Sci. Conf. **19**, 451–452.

Holsapple, K.A., Housen, K.R. (1986): Scaling laws for the catastrophic collisions of asteroids. Mem. S.A. It. **57**, 65–86.

Housen, K., Holsapple, K. (1990): On the fragmentation of Asteroids and Planetary Satellites. Icarus **84**, 226–253.

Housen, K.R., Schmidt R.M., Holsapple, K.A. (1991): 'Laboratory simulations of large scale fragmentation events. Icarus **94**, 180–190.

Kawakami, S., Mizutani, H., Takagi, Y., Kato, M., Kumazawa, M. (1983): Impact experiments on ice. J. Geophys. Res. **88**, 5806–5814.

Lange, A., Ahrens, T.J. (1981): Fragmentation of ice by low-velocity impact. Lunar Planet. Sci. Conf. **12**, 1667–1687.

Martelli, G., Rothwell, P., Giblin, I., Smith, P.N., Di Martino, M., Farinella P. (1983): Fragment jets from catastrophic break-up events and the formation of asteroid binaries and families. Astron. Astrophys. **271**, 315–318.

Marzari, F., Davis, D.R., Vanzani, V. (1995): Collisional evolution of asteroid families. Icarus **113**, 168–187.

Matsui, T., Schultz, P.H. (1984): On the brittle-ductile behavior of iron meteorites: New experimental constraints. J. Geophys. Res. Suppl. **89**, C323–C328.

Matsui, T., Waza, T., Kani, K. (1984): Destruction of rocks by low velocity impact and its implications for accretion and fragmentation processes of planetesimals. J. Geophys. Res. Suppl. **89**, B700–706.

Matsui, T., Waza, T., Kani, K., Suzuki, S. (1982): Laboratory simulation of planetesimal collisions. J. Geophys. Res. **87**, 10968–10982.

Melosh, H.J. (1989): *Impact Cratering: A Geologic Process*, Oxford Univ. Press.

Melosh, H.J., Ryan, E.V., Asphaug, E. (1992): Dynamic fragmentation in impacts: Hydrocode simulation of laboratory impacts. J. Geophys. Res. **97**, 14735–14759.

Nakamura, A., Fujiwara, A. (1991): Velocity distribution of fragments formed in a simulated collisional disruption. Icarus **92**, 132–146.

Nakamura, A., Suguiyama, K., Fujiwara, A. (1992): Velocity and spin of fragments from impact disruptions. I. An experimental approach to a general law between mass and velocity. Icarus **100**, 127–135.

Paolicchi, P., Cellino, A., Farinella, P., Zappalà, V. (1989): A semiempirical model of catastrophic breakup processes. Icarus **77**, 1187–1212.

Petit, J.-M., Farinella, P. (1983): Modelling the outcomes of high-velocity impacts between small solar system bodies. Celestial Mechanics and Dynamical Astronomy **57**, 1–28.

Ryan, E.V. (1992): *Catastrophic collisions: Laboratory impact experiments, hydrocode simulations, and the scaling problem*, Ph.D. dissertation, University of Arizona, Tucson.

Ryan, E.V., Davis, D.R. (1994): Energy partitioning in catastrophic collisions. Lunar Planet. Sci. Conf. **25**, 1175–1176.

Ryan, E.V., Davis, D.R. (1994): Asteroid collisions: The impact disruption of cooled iron meteorites. BAAS **26**, 1180.

Ryan, E.V., Hartmann, W.K., Davis, D.R. (1991): Impact Experiments 3: Catastrophic Fragmentation of Aggregate Targets and the Relation to Asteroids. Icarus **94**, 283–298.

Takagi, Y., Mizutani, H., Kawakami, S. (1984): Impact fragmentation experiments of basalts and pyrophyllites. Icarus **59**, 462–477.

Verlicchi, A., La Spina, A., Paolicchi, P., Cellino, A. (1994): The interpretation of laboratory experiments in the framework of an improved semi-empirical model. Planet. Space Sci. **42**, 1031–1042.

Signatures of Impacts in Quartz (Microstructures and Formation Mechanisms)

Jean-Claude Doukhan

Laboratoire Structure et Propriétés de l'Etat Solide (UA CNRS 234)
Université Sciences et Technologies de Lille
F-59655 Villeneuve d'Ascq-Cedex - France

Signatures d'impact dans le quartz (Microstructures et mécanismes de formation)

Résumé. L'impact de météorites frappant la surface de la Terre à grande vitesse engendre une compression dynamique qui peut atteindre plusieurs dizaines de GPa pendant une durée très courte (10^{-6} s à 1 s selon la taille de l'impacteur). Cette compression est accompagnée d'une forte augmentation de température qui peut induire des transitions de phase vers des polymorphes de haute pression et de la fusion locale ou même de la vaporisation. Des défauts spécifiques sont aussi produits dans les minéraux choqués. Ces derniers sont souvent considérés comme des signatures non ambiguës d'impacts météoritiques. Dans le quartz, ces défauts spécifiques sont de fines lamelles de macles mécaniques et des défauts plans appelés en anglais "planar deformation features" (PDF). Il s'agit de bandes très minces et rectilignes de silice amophe dense.

Abstract. The impact on the Earth surface of high velocity meteorite generates a high dynamic compression (up to several tens of GPa) during a very short time (10^{-6} to 1 s depending on the size of the impactor). It is accompanied by a severe temperature increase which can induce phase transitions toward high pressure polymorphs and local melting or even vaporization. Specific lattice defects are also produced in the shocked minerals which are considered as impact signatures. In quartz such specific defects are thin lamellae of mechanical twins and planar deformation features (PDF). These later ones are very straight and narrow bands of dense amorphous silica.

1 Introduction

It is only recently that impact cratering was recognized as an important geologic process on the Earth and other terrestrial planets (review in Melosh 1989). Actually more than 130 meteorite craters have been characterized on the Earth surface (Grieve 1991) but there must have been much more meteorite impacts, especially during the first billion years of the Earth history. The oldest ones are, however, no more detectable because of the dynamic behavior of the Earth which leads to the subduction of the older parts of the plates. Among the very large and very old impact craters, Sudbury in Canada is the largest structure (approximately 200 km) and Vredefort (South Africa) is the oldest one (approximately two billions years). The idea that impact cratering deeply affected the evolution of the Earth, including biologic evolution, is now well accepted. The last major biologic crisis occurred 65 millions years ago (Cretaceous/Tertiary boundary or KTB). Alvarez et al. (1980) suggested that it must result from the impact of a giant meteorite. This scenario was (and still is) challenged by other authors who believe that mass extinction was more gradual than what is expected from a meteorite impact. These authors rather believe that the crisis results from the huge lava flows called Deccan traps which occurred in the south of India over a period of time of approximately one million years centered on the K/T boundary (Courtillot, 1988). A number of impact signatures were contested, in particular the specific defects observed in quartz and called PDFs. Officer and Carter (1991) claimed they detected similar defects in structures clearly not related to meteorite impacts. Such a controversy was possible because the key defects were unsufficiently characterized (observations by optical microscopy only). This underlines the need for more complete characterizations by using transmission electron microscopy (TEM) and detailed comparison of defect microstructures produced in well characterized impacts with those produced by shock recovery experiments in known shock conditions. These detailed studies have been undertaken and yielded some clear information. For instance, among the relevant questions to be answered, it was necessary to know if these defects are identical in all shocked materials or if there are several types, some being specific of shock deformation (they would be the non ambiguous impact signatures) while others could be produced by a variety of causes and could not be considered as impact signatures. We review here these recent results.

It is well known that all terrestrial rocks suffered tectonic deformation. Shock-induced defects thus superimposed on previous defects produced by tectonic deformation. In addition the old crater structures suffered post-shock metamorphism which may have affected (and possibly partially overprinted) the original shock-defects. It appears necessary for a comprehensive presentation to start with a brief overview of the lattice defects produced by tectonic processes (first section). The next section presents recent TEM observations on shocked quartz, mainly from impact sites with various ages. The formation mechanisms of these shock-defects are discussed in the last section.

2 Lattice Defects Produced by Tectonic Processes

Almost all minerals of the crust and the upper mantle are crystalline and the response of almost all crystals to an applied stress is plastic strain. At low strain rate and moderate temperature (this is typically the case of plate tectonics) plastic strain occurs by the motion and multiplication of lattice defects called dislocations (Hirth and Lothe 1971). They are linear elastic singularities characterised by their Burgers vector noted $\mathbf{b}$ which measures the intensity of the defect. Figure 1 is a schematic representation of a straight dislocation line in the middle of a simple cubic crystal. The dislocation line is seen edge-on, it bounds an extra half plane, the intensity of the defect is measured by the thickness of this extra slab. A dislocation is a specific atomic configuration to which an elastic energy is associated. One demonstrates that this elastic energy is distributed in the whole crystal and for a unit length of dislocation line this energy is proportional to $\mu \mathbf{b}^2/4\pi$ where μ is the elastic shear modulus of the crystal. Therefore only dislocations with the smallest **b** exist and the smallest **b** are the base vectors of the primitive cell of the crystal lattice considered. Two types of dislocation motion are possible :

1. Glide. It requires the breaking of the interatomic bonds labeled AB on Figure 1 followed by the reconstruction of the bonds AC. This is the only possible motion of the dislocation at low temperature.
2. Climb. Vacancies (i.e. missing atoms) migrate toward the dislocation core in such a way that the extra half plane shortens (climbs). It is clear that this motion requires a minimum density of vacancies with a non vanishing mobility. This situation occurs only at high temperature. Therefore both, glide and climb motions are possible at high T but climb is generally much slower than glide because the equilibrium density of vacancies and their mobility are quite low, even at the higher temperatures (typical orders of magnitude are : atomic vacancy concentration $\approx 10^{-6}$; vacancy mobility $\approx 10^{-12}$ m.s^{-1}).

There is a formal analogy between the rate of plastic deformation and the rate of electric transport. The mobile dislocations are equivalent to electric carriers, the Burgers vectors are the elementary "charges" transported and the "current" is the product of the elementary charge (**b**) times the carrier density (density of mobile dislocations) times their mobility. This mobility depends on both the temperature and the external solicitation (i.e. the applied stress component in the glide plane, in the glide direction). When plastic strain stops, many dislocations still are present in the crystal. Their topology is a snapshot of the carrier density. At low temperature the mobility of dislocations is quite small and a given strain rate can be achieved only with a large dislocation density. In addition all the dislocations are confined in their glide planes which are in almost all crystal structures low index crystallographic planes containing the Burgers vector. In contrast, at high temperature, both glide and climb occur. Most of the plastic strain is achieved by dislocation glide, but dislocation climb allows some reorganization of the dislocation topology to occur in order to minimize the elastic energy stored in the crystal. One shows in standard text books on dislocations and mechanical properties of solids that dislocation organizations minimizing the elastic energy form bidimensional defects called subgrain boundaries (SGB) where dislocations are parallel and regularly spaced.

The numerous minerals occurring in the crust have different mechanical properties which can be interpreted by the specific properties of their dislocations (mobility, choice of glide planes, ...). We illustrate this brief review with the case of quartz which is an abundant and ubiquitous mineral of the Earth's crust. Quartz is the stable phase at ambient conditions of the compound SiO_2. In contrast with many other minerals which are solid solutions, the composition of

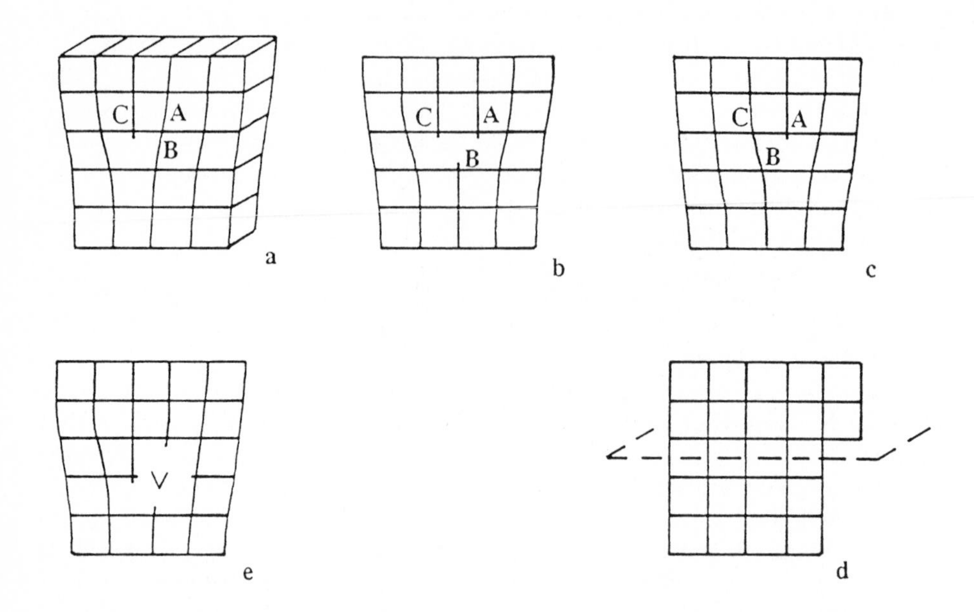

Fig. 1. Simple cubic crystal lattice with an edge dislocation in its middle.

natural quartz grains is always very close to pure silica, i.e. quartz presents a very low solubility for all the elements of the periodic table. The main impurities are on the one hand the substitution of Si^{4+} by Al^{3+} or Fe^{3+} ions associated with monovalent ions Na^+ or Li^+ (for preserving the electroneutrality) and on the other hand water (one Si^{4+} substituted by $4H^+$ or one SiO_2 by $2H_2O$). The maximum concentration of any of these impurities never exceeds some tens to one hundred atomic parts per million (at. ppm). In most crystals no strong impurity effects are expected from such small concentrations and an intrinsic (independent of impurity content) thermomechanical behavior is thus expected for quartz. It is remarkable, however, that very small amounts of water dissolved in the quartz lattice dramatically affect a number of its physical properties. For instance the piezoelectric performances of quartz resonators widely used in electronic industry (high precision clocks and time bases, electronic filters, ..) are dramatically decreased by the presence of very small amounts of water. Crystal growers have learnt to synthesize very dry crystals with water contents as low as $[H]/[Si] \approx 20$ at. ppm or even less (the actual detection limit of the infrared spectrometry technique used for detecting water in quartz is of the order of 20 ppm). Along the same lines, the mechanical strength of quartz dramatically decreases if the crystal contains water amounts $[H]/[Si] \geq 100$ ppm. Dry quartz is extremely strong and would not be ductile in the deformation conditions occurring in the Earth crust. Fortunately water is pervasive in the crust and quartz grains are wet and thus highly ductile. The phenomenon of hydrolytic weakening of quartz was discovered by Griggs (1967). This solved a long lasting apparent contradiction between strong evidences that quartz is highly ductile in natural deformation conditions (at moderate stress and temperature) while many laboratory experiments performed on gem quality crystals (i.e. very dry ones) clearly showed a very strong and brittle behavior.

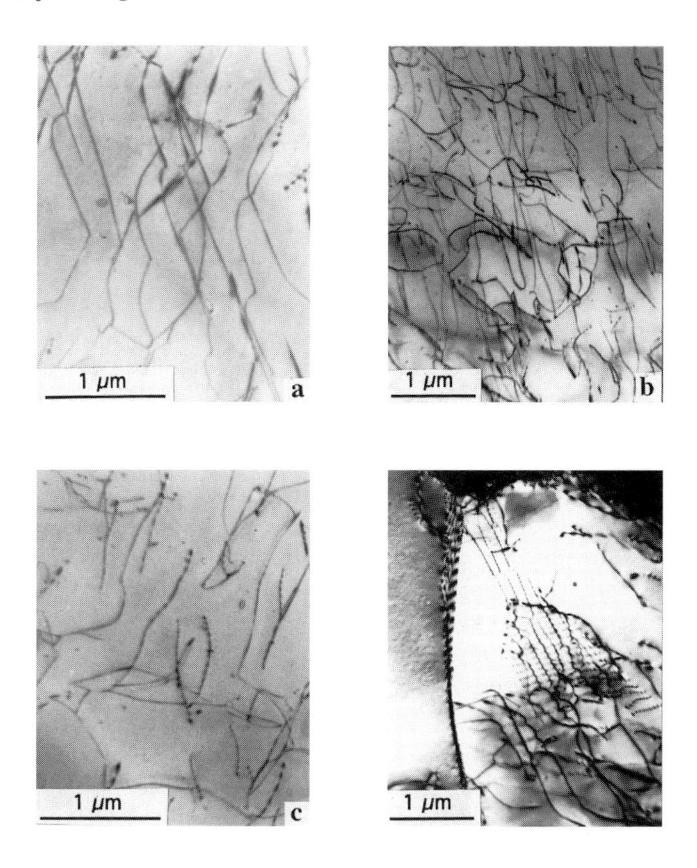

Fig. 2. Representative dislocation topologies in experimentally and naturally deformed wet quartz: a) Quartz experimentally deformed at $T = 690\ ^oC$ and $\dot{\varepsilon} \approx 10^{-6}\ s^{-1}$); b) Quartz experimentally deformed at $T = 500\ ^oC$, $\dot{\varepsilon} \approx 10^{-6}\ s^{-1}$); c) Quartz experimentally deformed at 900 ^oC; d) quartz naturally deformed.

Figure 2 is a representative series of TEM micrographs of wet quartz samples deformed at various temperatures and applied stresses. Figure 2a shows the defect microstructure in a very wet specimen ([H]/[Si] $\approx$ 1000 ppm) deformed at T = 690 ^oC and at a constant strain rate $\dot{\varepsilon} \approx 10^{-6}\ s^{-1}$. In these conditions quartz is weak and ductile with an elastic limit $\leq$ 100 MPa. The density of mobile dislocations necessary to produce this deformation rate is $\approx 10^{12}\ m/m^3$ or $10^{12}\ m^{-2}$. There is no place in this brief review for discussing the origin of the contrast of TEM micrographs. It is sufficient at this level to say that dislocations are imaged as dark lines on a clearer background in what is called bright field conditions or BF (dark field conditions are the complement of BF and dislocations thus appear as bright lines on a darker background). One also notices on Fig. 2a that the dislocation lines tend to be parallel to some preferred directions which have been characterized. They are the simple crystallographic directions **a**, **c**, and **a** ± **c** (**a** and **c** are the unit vectors of the primitive cell). These directions correspond to the ones of minimum dislocation mobilities. Figure 2b illustrates a deformation test at lower T and similar strain rate. The density of mobile dislocations necessary for accommodating the imposed strain rate is appreciably higher ($\approx 5 \times 10^{12}$ m^{-2}) because dislocation mobility decreases with decreasing T; a higher density of "deformation carriers" i.e. dislocations is thus necessary to achieve the same strain rate. One notices again one time that dislocations exhibit preferred orientations because of their mobility anisotropy. In contrast Fig. 2c corresponds to a deformation test at high temperature. The dislocation density is appreciably lower and their mobility anisotropy has disappeared. The dislocation lines are curved, no prefered orientations are detected. One observes from place to place a new dislocation topology consisting of regularly spaced dislocations forming two dimensional lattice defects called subgrain boundaries (SGB). SGBs result from efficient dislocation climb which allowed the minimization of the dislocation long range stress fields (i.e. minimization of their elastic energy). Finally Fig. 2d corresponds to a natural deformation at very low strain rate (tectonic processes are generally considerably slower than what can be done in laboratory experiments; natural strain rates may be as slow as $\approx 10^{-12}$ to 10^{-15} s^{-1}). The dislocation topology essentially consists of SGBs with few free dislocations (i.e. few isolated dislocations not organized in SGBs). One also observes tiny voids or bubbles. They result from water precipitation when the (T,P) conditions changed as the crystal was progressively brought to the Earth surface. Indeed thermodynamic computations show that the equilibrium solubility of water depends on T and P and practically cancels at ambient conditions. These bubbles have preferentially nucleated on preexisting lattice defects like SGBs because their energy of nucleation is lowered on such defects. The familiar experiment of bubble formation in water heated in a container depicts a similar situation. The tiny bubbles of vapor preferentially nucleate on the container surface because their nucleation is lowered at the surface contact.

3 Shock-Induced Lattice Defects in Quartz

TEM characterizations of the defects induced by meteorite impacts are recent and only a few minerals have been investigated in detail. TEM observations of shocked quartz have revealed a variety of shock-defects (Goltrant et al. 1991, 1992). This variability may, however, reflect the modifications due to post shock metamorphism. One has thus to take into account these possible changes when studying old impact structures like for instance the Vredefort structure (South Africa, ≈ 2 billions years, Engelhart and Bertch 1969). Other well known impact structures like Slate Islands or La Malbaie (both in Canada) are also quite old (approximately 300 millions years old, McIntire 1962; Robertson 1975). The Ries Cater in Germany is among the most recent large craters and furthermore it still has its blanket ejecta. This material which was never buried is probably the least affected by post-shock thermal events. It is thus ideal for the characterization of original shock defect microstructures. The detailed studies of Goltrant et al. (1991, 1992) on this material can be summarized as follows. The density of shock defects varies appreciably from sample to sample, probably because they suffered different shock intensities before being ejected (the impact generates a spheric shock wave, consequently the shock intensity decreases with increasing distance to the center of the impact). However, all exhibit the same types of shock-defects. In addition to the so-called shock-mosaicim patterns which are better visible in optical microscopy one observes in TEM pervasive thin mechanical twin lamellae in the basal plane and the so-called "planar deformation features" or PDF.

3.1 Mechanical Twinning

In contrast with slow dislocation motion which involves the breaking of bonds one after another ("civilian process"), mechanical twinning is a "military process" i.e. a process involving the cooperative motion of many atoms. The resulting twin lamella has the same crystal structure as the matrix but with a different orientation (Fig. 3a).

It is generally assumed that mechanical twinning results from the rapid and cooperative glide in adjacent planes of partial dislocations (dislocations with a Burgers vector smaller than a lattice repeat unit). Partial dislocations still are visible (arrows) in twin boundaries on the TEM micrograph Figure 3b. Theoretical computations of TEM contrast predict fringe patterns for the imaging of the boundaries between twins and matrix. Such fringes are clearly visible on Fig. 3b. As mechanical twinning is a shear strain, it is activated by a shear stress. Twinning can be activated by a shock wave only if its associated stress tensor develops a large enough deviatoric component in the glide (shear) plane, in the glide (partial Burgers vector) direction. Of course these thin lamellae of mechanical twins superimpose on the lattice defects left by the tectonic deformation which occurred before the shock and to those produced by post-shock metamorphism. These other "background" defects are free dislocations, dislocations organized in SGBs, some water bubbles and so on.

No such twin lamellae are observed in naturally deformed quartz or in quartz deformed in laboratory experiments at moderate strain rate. Shock recovery experiments on quartz samples do not reveal mechanical twinning either. The only known exception is an experiment performed by McLaren et al. (1967) on a dry quartz. Because this quartz specimen was dry, it had to be compressed under large confining pressure to prevent fracturing. A very high differential stress (maximum difference between the eigen values of the stress tensor) roughly estimated ≥ 2 GPa had to be applied for activating plasticity and a similar mechanical twinning was observed to occur in this experiment. There is no known data for the twinning threshold in wet quartz. However it is to be remembered that the hydrolytic weakening phenomenon in wet quartz is efficient only in situations where the water point defects have time enough for efficient migration (migration or diffusion is a very slow step by step process). Therefore hydrolytic weakening

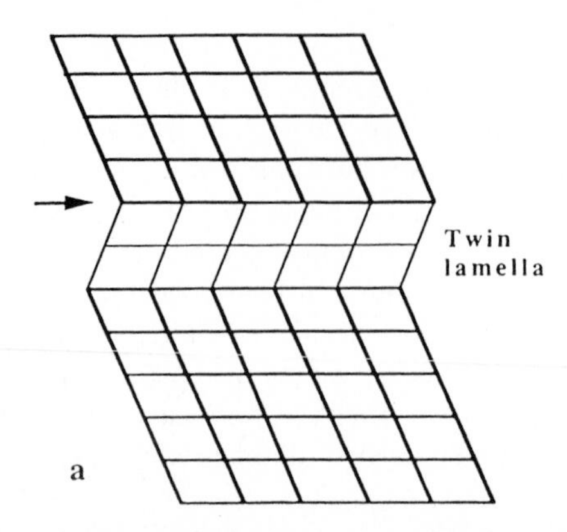

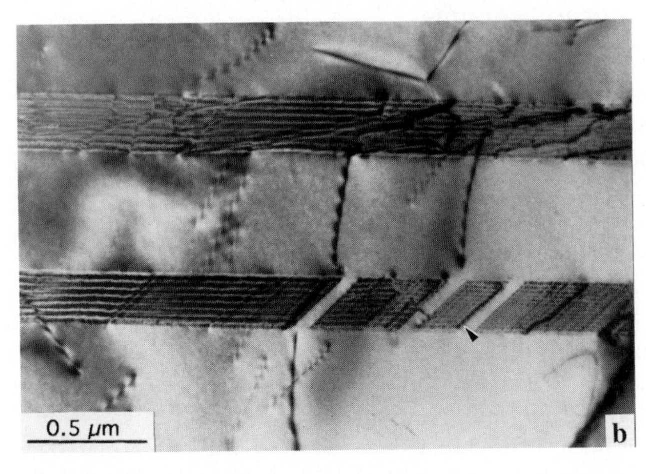

Fig. 3. Mechanical twinning in shocked quartz from the ejecta of the Ries Crater impact (Germany): a) Twinning mechanism by glide of partial dislocations. b) Inclined twin lamellae; their boundaries are underlined by a fringe pattern.

is probably totally inefficient in the case of shock deformation. The pervasive occurrence of thin mechanical twin lamellae in the basal plane of shocked quartz like the ones shown on Fig. 3b thus indicates that the shock wave which produced them was not perfectly spherical (hydrostatic conditions) and its associated differential stress was at least 2 GPa. This is a high value which is never reached in usual tectonic processes. Basal mechanical twin lamellae are thus an unambiguous signature of shock wave deformation, i.e. of meteorite impacts.

The recent TEM investigations of the collar of the Vredefort structure (South Africa) reveals the pervasive occurrence of similar mechanical twin lamellae. In this case the thin twin lamellae are decorated by tiny bubbles (< 100 nm) which are partially filled with a non characterized fluid (Fig. 4). These fluid bubbles must result from the slow diffusion of the water defects during post shock metamorphism. Water precipitation preferentially nucleated on defects (twin lamellae). This is the well known process of heterogeneous precipitation. The origin of this 2 billions years old geologic structure was a debated problem

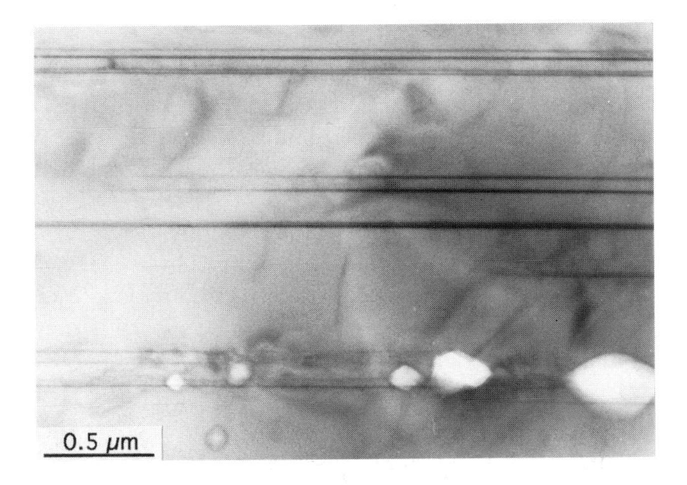

Fig. 4. Twin lamellae (seen edge-on) decorated with tiny water bubbles in a quartz grain from the collar of the Vredefort structure.

for years. A number of field indices suggested an impact origin altered by a more recent high grade metamorphism stage which would have partially hidden initial impact signatures. The recent TEM observations of mechanical twins thus prove that i) the Vredefort structure does result of a meteorite impact, ii) the mechanical twins are not overprinted by post shock metamorphism even by high grade metamorphism. Mechanical twin lamellae thus constitute very precious shock indices for old impact structures. Unfortunately they are not detectable optically if they are not decorated by tiny water bubbles. They can be visualized (and characterized) only by TEM which is a technique much more complex and time consuming that simple optical observations.

Detailed investigations have been performed on a few crater structures at various distances to their center. These studies have shown that twinning is rare in the vicinity of the impact center, but become progressively more abundant at increasing distances. Along the same lines, quartz experimentally shocked at peak pressures 20 to 40 GPa reveals no twinning (Gratz 1984; Gratz et al. 1988). The reason for these different behaviors probably results from different shock wave structures. The highly sophisticated assemblies actually used for shock recovery experiments have been designed in order to produce shock waves as plane as possible with stress tensors as spheric as possible (i.e. with no differential stress). In contrast meteorite impacts generate spheric waves which propagate over large distances and their stress tensor structures progressively change. The pressure intensity decreases but the differential stress increases due to the multiple reverberations and the complex boundary conditions controlling the propagation of the shock wave.

3.2 Planar Deformation Features

Since their discovery in 1962 by McIntire in the impact structure of Clear Water Lake (Canada), these very straight and very narrow defects have been considered by a number of authors as the most unambiguous indices of shock metamorphism. As they are easily detected optically (Fig. 5) they look very convenient indices for the rapid characterization of impact structures. Their fine structure, however, cannot be elucidated by optical microscopy only, and a number of terrestrial structures have been reported to exhibit quartz bearing PDFs although it was clear that they did not result from meteorite impacts. More detailed characterizations of PDFs microstructures are thus necessary if they are to be used as unambiguous signatures of meteorite impacts.

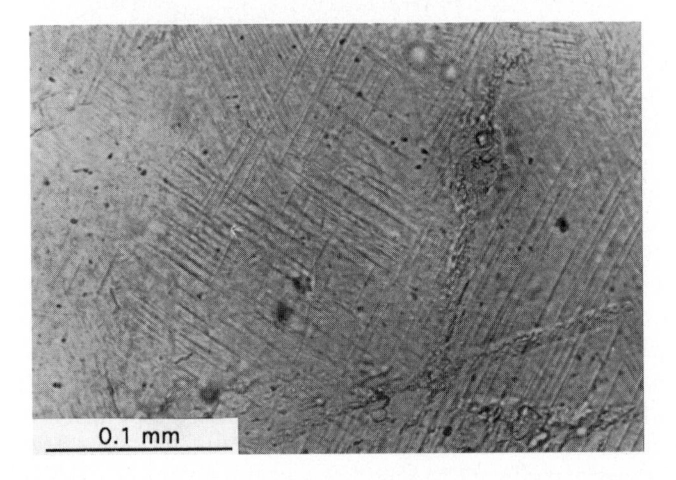

Fig. 5. Multiple sets of PDFs in quartz from the Ries Crater (optical micrograph, unpolarized light).

We mentioned above that the blanket ejecta of the Ries crater (Germany) appears the best material for this investigation because the structure is relatively young (15 millions years) and its ejecta suffered a minimum post shock metamorphism. Figure 6 is a TEM dark field micrograph showing PDF microstructures in this shocked quartz. A dark field image is formed with the beams diffracted at an angle close to (but not identical to) the diffraction Bragg angle. Amorphous material generates scattered electron beams in all directions (no Bragg angles) and thus brings a higher contribution than a crystalline material which diffracts the electronic waves only at specific angles (Bragg angles). In dark field amorphous material thus appears brighter than crystalline material. This technique allows the detailed characterization of PDFs. They are straight and narrow lamellae of a mixture of amorphous silica and very tiny crystallites of quartz (size $\leq$ 10 nm). PDF thickness never exceeds 100 nm and the lamellae extend over the whole grains parallel to a few crystal planes with crystallographic indices (the so-called Miller-Bravais indices) $\{10\bar{1}n\}$ with n = 1, 2, 3, or 4. n = 2 and 3 are the most frequent values. These planes are illustrated on Fig. 7 within the unit hexagonal cell of quartz. Of course, these PDFs superimpose on preexisting lattice defects resulting from the tectonic deformation prior to the impact. These preexisting defects are free dislocations, SGBs, some water bubbles and so on. This background is similar to what is observed on Fig. 2d.

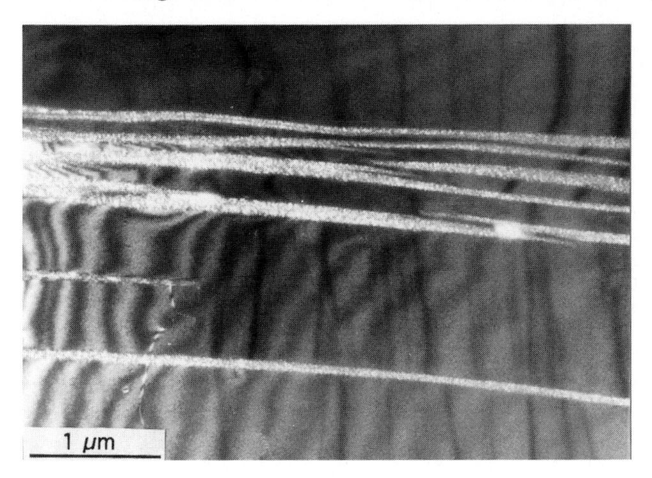

Fig. 6. TEM micrograph (dark field) of PDFs in quartz from the ejecta of the Ries Crater.

In old impact structures like Clear Water Lake or Manson (Iowa, USA, 65 million years old) one also observes in TEM (in addition to the usual background of lattice defects left by tectonic deformation) straight and narrow contrasts parallel to the same $\{10\bar{1}n\}$ planes. In contrast with the case of Ries Crater, in these rocks the fine structure of these defects consists of a high density of entangled dislocations with many tiny fluid bubbles (Fig. 8). These PDFs have no more their original structure. What is actually observed must result from post shock alteration and metamorphic processes. One can suggest the following scenario. Like for the case of Ries Crater, the original PDFs consisted of thin lamellae of amorphous silica in a quartz matrix. Like for all other quartz in the Earth crust this quartz matrix was wet. Amorphous silica is known to be able to dissolve considerably larger amounts of water than crystalline quartz (up to 6% to be compared to 100 to 1000 ppm for the crystalline phase). During post shock thermal events, the water defects migrated from the surrounding crystalline matrix toward the amorphous bands, the water content of which increased. When the amount of water became large enough, amorphous silica started to recrystallize, a process which is known to occur when it is efficiently assisted by large amounts of water. The newly formed quartz crystal was clearly supersaturated in water and a concomitant precipitation of the excess of water occurred in the form of tiny water bubbles in the areas which had been initially PDFs. In addition many dislocations nucleated and grew in order to accommodate the lattice distortions around the bubbles.

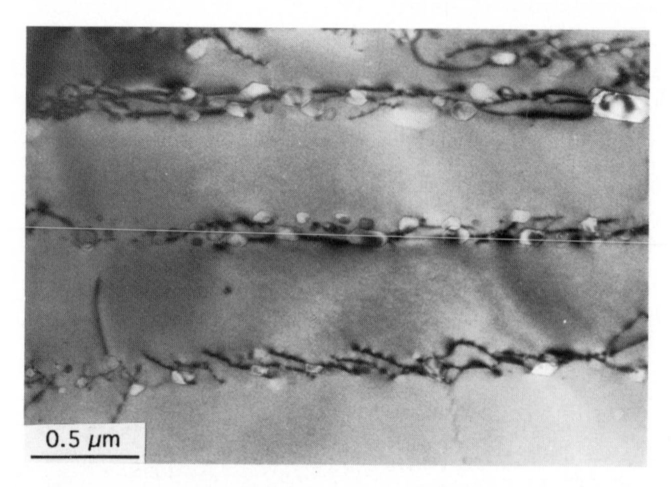

Fig. 7. TEM micrograph of altered PDFs of the Manson impact (entangled dislocations with tiny water bubbles connected to them).

Finally it has to be mentioned that, at the high magnification of TEM micrographs, all the other pseudo PDF structures and similar contrasts which have been reported to occur (from optical observations only) in terrestrial structures (i.e. which do not result from meteorite impacts like land sliding at Köfels, Austria, explosive volcanism at Toba, Sumatra, ...) markedly differ from what is shown on Figs. 6 and 8. These defects are not confined in $\{10\bar{1}n\}$ planes, they are not straight and narrow contrasts, they are not amorphous silica, ... Briefly one can say that they are not PDFs, and consequently they do not result from meteorite impacts. PDFs thus appear unambiguous signatures of meteorite impacts on the Earth, although their fine structure is less resistant to post shock metamorphism than mechanical twins (for instance PDFs have completely disappeared in the Vredefort structure). Again one time the use of optical microscopy only can lead to erroneous diagnostics. Complete TEM characterization is necessary for deciding if some structures observed at the optical microscope are or are not PDFs.

4 A Formation Model for PDFs

The formation of thin twin lamellae during meteorite impacting is easily understood. Twinning is known to be a very rapid shearing process occurring under large applied differential stresses. It can thus be induced by a shock wave, as soon as its differential stress overcomes a threshold which, in the absence of other data, is presumed to be ≥ 2 GPa. The formation of PDFs is less straightforward. One has to interpret two important observations: i) a phase transition occurs (crystalline quartz $\rightarrow$ amorphous silica), and ii) the pieces of the new phase have a very peculiar shape; they are very straight and narrow lamellae extending over the whole grains, their thickness is ≤ 100 nm and their orienta-

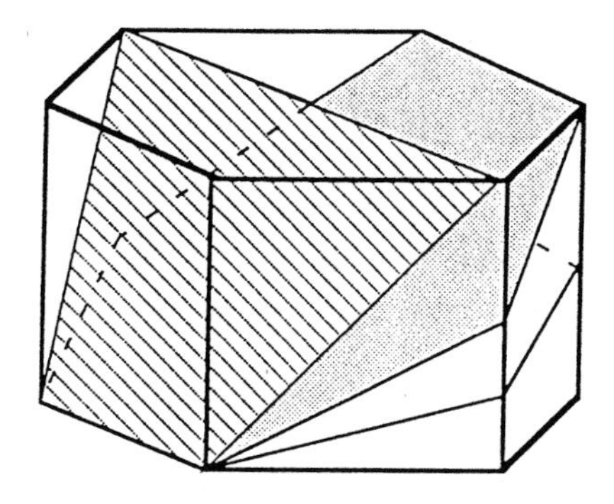

Fig. 8. $\{10\bar{1}n\}$ planes shown in the hexagonal unit cell of α-quartz.

tions parallel the $\{10\bar{1}n\}$ planes. In addition the few TEM investigations performed on experimentally shocked (and recovered) quartz (Gratz 1984; Gratz et al. 1988) indicate that the amorphous PDF lamellae only form in those of the $\{10\bar{1}n\}$ planes which are parallel or approximately parallel to the shock wave direction. We thus have to interpret both phenomena, phase transition toward a metastable phase (amorphization) and the specific shapes of the products of this transformation.

4.1 Amorphization

Amorphization of quartz (and other silicates) under static pressure at room temperature has been the subject of a number of recent experimental studies in diamond anvil cells. This technique generates very large and roughly hydrostatic pressures on very small samples and these experiments have clearly demonstrated that quartz becomes amorphous above a pressure $\approx$ 10 GPa at room temperature (Hemley 1988; Cordier et al. 1993). At higher temperature it would probably transform into stishovite which is the stable crystalline phase at high P and T, but the reorganization of the various atoms into a crystal (i.e. a completely ordered phase) is prohibited by the low temperature of the experiments. An amorphous phase with a higher compressibility and a lower specific volume than crystalline quartz is thus the only possible (metastable) phase which can be produced by high compression at moderate temperature. This amorphous phase cannot have the same internal structure (network) as usual glass obtained by high temperature melting of quartz. Indeed this usual glass has a specific mass markedly lower than the one of crystalline quartz ($\approx$ 2.2 gcm^{-3} versus 2.6 for quartz). Crystalline quartz as well as usual silica glass are constituted of a network of corner linked SiO_4 tetrahedra. The dense amorphous silica phase in the PDFs must contain an appreciable amount of corner sharing SiO_6 octahedra in strong similarity with the dense stishovite crystalline phase. Indeed transforming a network of SiO_4 tetrahedra into a network of SiO_6 octahedra leads to a volume compaction of $\approx$ 40%. These data may explain why quartz submitted to the high pressure of a shock wave partially transforms into a dense amorphous phase but this does not explains why the nuclei of the transformed phase have a very specific shape. In static compression experiments the amount of dense amorphous phase is observed to increase with increasing pressure and the transformation is completed for P $\approx$ 40 to 50 GPa, but microscopic investigations show that the domains of the newly formed amorphous phase have no particular shape. The very specific shape of the PDFs induced by shock compression must be related to the dynamic aspect of the shock wave (mobile shock front, short duration of the compression stage favoring nucleation but preventing growth of the amorphous domains). We also have to understand the mechanism responsible for preferential nucleation in $\{10\bar{1}n\}$ planes. The stability criterion proposed by Born (1939, 1954) is useful for modeling PDF formation. This criterion states that a crystalline phase is no longer stable if its elastic moduli become negative (the response to a compressive stress would become an elongation). We can thus compute in anisotropic elasticity the shear moduli of the $\{10\bar{1}n\}$ planes for various confining pressures. This is a straightforward (but tedious) calculation which just requires the knowledge of the variation of the elastic constants versus P. The basic law of elasticity (Hooke's law) states that an applied stress tensor σ produces an elastic deformation tensor ε related to σ by the linear equation

$$\sigma = \mathbf{C} : \varepsilon \quad \text{or} \quad \varepsilon = \mathbf{S} : \sigma \tag{1}$$

C and **S** are the fourth rank tensors of elastic stiffness and elastic compliance respectively ($\mathbf{S} = \mathbf{C}^{-1}$). The need of precise data for manufacturing good quartz resonators has led to precise measurements of the **C** components at various T, but no measurements have been performed at high P so far. Fortunately these data were recently computed from the first principles of quantum mechanics (Purton et al. 1993) and the results fit very well the low pressure experimental data. An elastic shear modulus is related to a system constituted of a plane (shear plane) with normal **n** and an associated direction **d** lying in this plane (shear direction). The shear modulus is the ratio of the shear stress on (**n**,**d**) divided by the corresponding shear strain. Taking advantage of the symmetries of the stress and strain tensors ($\sigma_{ij} = \sigma_{ji}$ and $\varepsilon_{ij} = \varepsilon_{ji}$) one may represent them by pseudo vectors $[\sigma_J]$ and $[\gamma_J]$ in a 6 dimensions space by using the following equivalence relations for the indices :

$$11 \Rightarrow 1;\ 22 \Rightarrow 2;\ 33 \Rightarrow 3;\ 23 \Rightarrow 4;\ 31 \Rightarrow 5 \text{ and } 12 \Rightarrow 6$$
for the stress tensor
$$\varepsilon_{ij} \Rightarrow \gamma_J \text{ for } J = 1, 2, \text{ and } 3;\ 2\ \varepsilon_{ij} \Rightarrow \gamma_J \text{ for } J = 4, 5, \text{ and } 6 \tag{2}$$
for the strain tensor.

The Hooke's law is now written

$$\sigma_I = \sum C_{IJ} \gamma_J \tag{1'}$$

and $[C_{IJ}]$ is a 6×6 symmetric matrix representing the fourth rank stiffness tensor **C**.

For α-quartz, the Born criterion leads to the following general relations :

$$C_{11} - C_{12} > 0 \;\; ; \;\; (C_{11} + C_{12})C_{44} - 2C_{13}^2 > 0$$

$$C_{11}C_{33}C_{44} - C_{13}^2 C_{44} - C_{14}^2 C_{33} > 0 \tag{3}$$

As the C_{IJ}'s are pressure dependent, these relations allow, in principle, the critical pressure at which quartz is no more mechanically stable to be determined. They do not provide, however, information about the particular planes onto which an instability will occur. To determine these particular planes, we have to compute the pressure dependence of the shear moduli for the (**n**,**d**) relevant shear systems i.e. for the $\{10\bar{1}n\}$ planes and various **d** directions in these planes. Figure 7 shows the orientations of these planes drawn in the unit hexagonal cell of α-quartz.

Let us consider a simple situation where the applied stress tensor has only one non vanishing component acting on the required (**n**,**d**) system and, considering the symmetry of quartz, the strain tensor is given by the following equation (Hooke's law) with in addition the relation $S_{66} = 2(S_{11} - S_{12})$

$$\begin{bmatrix} \gamma_1 \\ \gamma_2 \\ \gamma_3 \\ \gamma_4 \\ \gamma_5 \\ \gamma_6 \end{bmatrix} = \begin{bmatrix} S_{11} & S_{12} & S_{13} & S_{14} & 0 & 0 \\ S_{12} & S_{11} & S_{13} & -S_{14} & 0 & 0 \\ S_{13} & S_{13} & S_{33} & 0 & 0 & 0 \\ S_{14} & -S_{14} & 0 & S_{44} & 0 & 0 \\ 0 & 0 & 0 & 0 & S_{44} & 2S_{14} \\ 0 & 0 & 0 & 0 & 2S_{14} & S_{66} \end{bmatrix} \cdot \begin{bmatrix} 0 \\ 0 \\ 0 \\ 0 \\ 0 \\ \sigma_6 \end{bmatrix} \tag{4}$$

The shear modulus is $1/S_{66}$ for shearing in direction **d** // OX, and in the plane with normal **n** // OY. This yields the following result

$$\frac{1}{S_{66}} = \frac{C_{66}.C_{44} - C_{14}^2}{C_{44}} = \frac{(C_{11} - C_{12})C_{44} - 2C_{14}^2}{2C_{44}} \tag{5}$$

(it is to be remembered that the compact notation implies the following relations for the S_{IJ} coefficients: if both I and $J \leq 3$, $S_{IJ} = S_{ijkl}$; if one of these indices $= 4$ to 6, the other still being ≤ 3, $S_{IJ} = 2S_{ijkl}$ and if both I and $J = 4$ to 6, $S_{IJ} = 4S_{ijkl}$).

The shear modulus for the same direction **d** // OX but for $\{10\bar{1}n\}$ planes is obtained by rotating the XOZ plane by an angle α around the OX axis. The new elastic coefficients (referred to the rotated axes $OXY'Z'$) are C'_{66} and $1/S'_{66}$. As **C** and **S** are fourth rank tensors, the following relations hold for the rotation operation considered

$$\left.\begin{aligned} S'_{ijkl} &= \sum T_{im}.T_{jn}.T_{ko}.T_{lp}.S_{mnop} \\ \text{or} \quad C'_{ijkl} &= \sum T_{im}.T_{jn}.T_{ko}.T_{lp}.C_{mnop} \end{aligned}\right\} \tag{7}$$

T_{im} being the components of the rotation matrix (angle α around the OX axis)

$$[T_{im}] = \begin{bmatrix} 1 & 0 & 0 \\ 0 & \cos\alpha & \sin\alpha \\ 0 & -\sin\alpha & \cos\alpha \end{bmatrix} \tag{8}$$

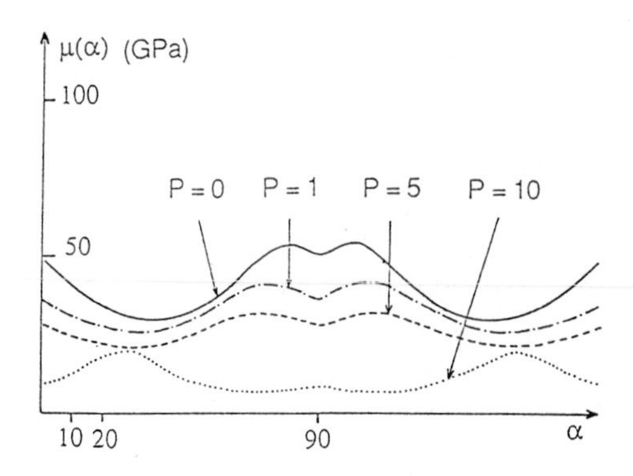

Fig. 9. Variation with pressure of the shear modulus in $\{10\bar{1}n\}$ planes.

and the shear modulus results

$$\mu(\alpha) = \frac{1}{S'_{66}} = \frac{1}{S_{66}\cos^2\alpha + S_{44}\sin^2\alpha + 2S_{14}\sin 2\alpha}$$
$$= \frac{(C_{11} - C_{12})C_{44} - 2C_{14}^2}{2C_{44}\cos^2\alpha + (C_{11} - C_{12})\sin^2\alpha - 2C_{14}\sin 2\alpha} \quad (9)$$

Its variation with pressure for various α values is shown on Fig. 9. One observes that μ is positive up to $P \approx 11$ GPa, whatever α. For larger P values $\mu(\alpha)$ becomes unstable in a few planes which correspond to the experimentally observed PDF planes. Therefore this simple elastic calculation explains why amorphization starts in $\{10\bar{1}n\}$ planes with $n = 1$, 2, 3, and 4. This elastic model does not explain, however, why further growth of PDFs is restricted to these planes (the amorphous nuclei could grow isotropically leading to roughly spheric amorphous domains as observed in samples statically compressed).

The dynamic aspect of the shock wave may provide a coherent model for the last stage of PDF formation i.e. for their anisotropic growth leading to the observed straight and narrow lamellae in the $\{10\bar{1}n\}$ planes. Figure 10 is a tentative interpretation of this last stage. The left side is the compressed zone. It is separated from the undisturbed zone (right side) by the moving shock front. Severe mismatch in this narrow shock front zone induces a severe elastic deformation gradient to which a large elastic energy is associated. This energy can be partially relaxed if small domains of the dense amorphous phase are nucleated in the shock front zone (in $\{10\bar{1}n\}$ planes). Their growth can then be triggered by the shock front as this later one propagates (see details of the model in Goltrant et al. 1992). This growth model explains why PDFs form narrow lamellae in the $\{10\bar{1}n\}$ planes approximately parallel to the propagation direction of the shock wave. It also implies that only dynamic compression (i.e. shock waves generated by meteorites impacts) can produce these typical defects which thus are unambiguous signatures of meteorite impacts.

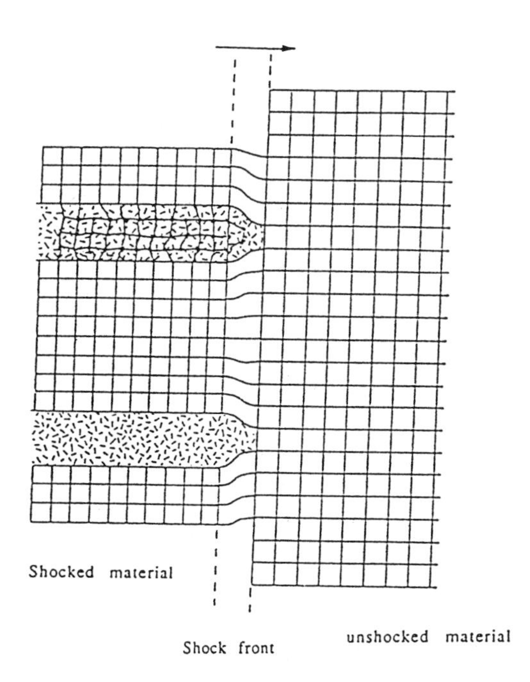

Fig. 10. Schematic model of the formation of PDFs.

References

Alvarez L.W., Alvarez W., Asaro F., Michel H.W. (1980): Extraterrestrial cause for the Cretaceous-Tertiary extinction. Science **208**, 1095-1108

Born M. (1939): Thermodynamics of crystals and melting. J. Chem. Phys. **7**, 591-603

Born M., Huang K. (1954): *Dynamical theory of crystal lattices* (Oxford University Press)

Cordier P., Doukhan J.C., Peyronneau J. (1993): Structural transformations of quartz and berlinite $AlPO_4$ at high pressure and room temperature. A transmission electron microscopy study. Phys. Chem. Minerals **20**, 176-189

Courtillot V. (1988): Deccan flood basalts and the Cretaceous/Tertiary boundary. Nature **333**, 843-846

Engelhart W. von, Bertsch W. (1969): Shock induced planar deformation structures in quartz from the Ries Crater, Germany. Contrib. Mineral Petrol. **20**, 203-234

Goltrant O., Cordier P., Doukhan J.C. (1991): Planar deformation features in shocked quartz; a transmission electron microscopy investigation. Earth Planet. Science Letters **106**, 103-155

Goltrant O., Doukhan J.C., Cordier P. (1992): Formation mechanisms of planar deformation features in naturally shocked quartz. Phys. Earth Planet. Int. **74**, 219-240

Gratz A. (1984): Deformation in laboratory-shocked quartz. J. Non Crystal Sol. **67**, 543-558

Gratz A.J., Tyburczy J., Christie J.M., Ahrens T., Pongratz P. (1988): Shock metamorphism of deformed quartz. Phys. Chem. Minerals **16**, 221-233

Grieve R.A.F. (1991): Terrestrial impact: the record of the rocks. Meteoritics **26**, 175-194

Griggs D.T. (1967): Hydrolytic weakening in quartz and other silicates. Geophys. J. Royal Astron. Soc. **14**, 19-31

Hemley J..R, Jephcoat A.P., Mao H.K., Ming L.C., Manghnani M.H. (1988): Pressure induced amorphization of crystalline silica. Nature **334**, 52-54

Hirth J.P., Lothe J. (1971): *Theory of dislocations* (Mc Graw-Hill)

McIntire D.B. (1962): Impact metamorphism at Clear Water Lake, Quebec. J. Geophys. Res. **67**, 1647-1653

McLaren A.C., Retchford J.A., Griggs D.T., Christie J.M. (1967): TEM study of Brazil twins and dislocations experimentally produced in natural quartz. Phys. Stat. Sol. **19**, 631-644

Melosh H.J. (1989): *Impact cratering, a geologic process* (Oxford University Press)

Officer C., Carter N.L. (1991): A review of the structure, petrology, and dynamic deformation characteristics of some enigmatic terrestrial structures. Earth Science Rev. **30**, 1-49

Purton J., Jones R., Catlow C.R., Leslie M. (1993): Ab initio potentials for the calculation of the dynamical and elastic properties of α-quartz. Phys. Chem. Minerals **19**, 392-400

Robertson P.B. (1975): Zones of shock metamorphism at the Charlevoix impact structure, Quebec. Geol. Soc. Amer. Bull. **86**, 1630-1638

III

Terrestrial Impacts

Dating of Impact Events

Urs Schärer

Laboratoire de Géochronologie, Université Paris 7 et IPG-Paris
2 place Jussieu, tour 24-25, 1e étage, F-75251 Paris cedex 05, France

Datation des impacts

Résumé. Ce chapitre présente les différentes méthodes utilisées, ou susceptibles d'être utilisées, pour déterminer l'âge d'un évènement d'impact sur les planètes terrestres ou les corps parents des météorites. Il existe deux approches qui sont : (1) Les méthodes géologiques telle que la stratigraphie qui est l'étude des successions d'événements géologiques, et (2) les méthodes basées sur la radioactivité naturelle (datation radiométrique) ainsi que les mesures d'isotopes produits par le rayonnement cosmique dans l'espace ou à la surface de la terre. L'accent est mis sur les méthodes nucléaires car elles donnent des âges absolus, contrairement aux méthodes géologiques qui ne fournissent que des âges relatifs. Les effets du métamorphisme de choc sur les minéraux et les roches se présentent sous l'aspect d'un cratère. Les différentes transformations de phases induites, par le passage de l'onde de choc, sont typiques des régimes de haute pression (20-60 GPa) et ne peuvent pas être produites par les processus géologiques classiques tels que le métamorphisme régional ou le magmatisme. Par conséquent, les impacts produisent des roches qui sont bien distinctes des roches de la croûte continentale ou océanique, appelées "les impactites". Leur composition varie entre : des verres de fusion (tektites et sphérules), des verres diaplectiques (phases amorphes), des brèches à matrice entièrement ou partiellement vitreuse, et des fragments de roches inclus dans une matrice de fragments très fins. Pour déterminer un âge, il est nécessaire qu'un équilibre isotopique s'installe entre les différentes phases nouvellement produites. Ces conditions sont potentiellement présentes dans les verres, produits par les températures résiduelles après la décompression ($> 1000^o$ C). Les meilleurs candidats, à plus grande échelle, sont : les échantillons prélevés dans les couches de fusion du cratère, dans les matrices vitreuses des brèches de la région du cratère, ainsi que les tektites ou sphérules que l'on trouve dans les éjections lointaines. La présence de ce type d'éjections permet de dater des évènements d'impacts par l'intermédiaire des couches préservées dans les sédiments. Un excellent exemple pour ce scenario est la limite Crétacé-Tertiaire où une grande série de minéraux produits par l'impact ont été préservés. Dans certains cas, on peut aussi tracer des événements même si l'on ne connait pas le cratère. La majorité des roches produites par le métamorphisme de choc ne sont pas en équilibre isotopique, et souvent les valeurs mesurées représentent des âges de mélange où les systèmes perturbés ne donnent pas l'âge de l'impact. Seuls, la combinaison des différentes méthodes de datation, ainsi qu'un choix très sélectif des échantillons, permettent la datation des impacts.

Abstract. This chapter presents the different methods used or potentially useful to determine the age of an impact event on terrestrial planets or meteorite parent bodies. Two approaches exist: (1) geological methods such as Stratigraphy, which is the study of succeeding geological events, and (2) methods based on natural radioactivity (radiometric dating), as well as the measurement of isotopes produced by cosmic rays in space or on the Earth's surface. The nuclear methods are preferentially discussed because they yield absolute ages, whereas geological methods define relative ages only. Shock metamorphic effects on minerals and rocks are presented in the frame of an impact crater. The different phase transformations induced by shock wave passage are typical for high-pressure regimes (20-60 GPa) which cannot be produced by any classical geological process like regional metamorphism or magmatism. In consequence, impacts produce rocks called "impactites" that are very distinct from lithologies of continental or oceanic crust. Their composition varies between molten glass (tektites and spheruls), diaplectic glasses (amorphous phases), breccias with either entirely or partially glassy matrix, and rock fragments enclosed in a fine grained fragmental matrix. To determine an age, isotopic equilibrium must be achieved among the newly produced phases. Such conditions are potentially present in glasses that are produced by residual heat after decompression ($> 1000^o$ C). The best candidates on the large scale are: samples taken in melt layers of craters, in the glassy matrix of breccias in the crater area, as well as tektites and spheruls found in distant ejecta. The presence of such types of ejecta allows to date the impact event through the analyses of layers preserved in sediments. An excellent example for this scenario is the Cretaceous-Tertiary boundary where a large series of impact produced minerals have been preserved. In certain cases, it is possible to trace events even if the crater remains unknown. The majority of shock-wave produced rocks are not in isotopic equilibrium, and often the values measured represent mixed ages or disturbed systems, not giving the impact age. Only the combination of different dating methods, and very selective sample selection allows impact dating.

1 Introduction

In asking questions on time-frequency relationships and size distributions of hypervelocity collisions in the solar system, precise age dating of impact events is of fundamental importance. Although such dating has always been an essential part of meteorite and lunar research, its precision requirements became strikingly clear in the context of discussions on the Cretaceous-Tertiary (K/T) mass extinction, which may have been caused by a large impact event 65 m.y. ago. Compared to the geological time-scale, impacts are instantaneous events, allowing very precise dating with age-resolutions on the order of 1 m.y. To date impact events, two different approaches have to be distinguished :

(i) The determination of "relative" ages by studying successions of craters, as well as their relationships with other geological phenomena such as volcanic deposits and flows (e.g. on the Moon and Mars), erosion structures (e.g. channels on Mars), or sedimentary deposits (e.g. regolith on the Moon and Mars).
(ii) The determination of "absolute", i.e. radiometric ages by using naturally occurring nuclear phenomena in rocks and minerals. This latter approach also includes the production and decay of isotopes produced by radiation either in space (spallation) or on Earth (terrestrial exposure ages).

Potentially, the determination of ages is possible on meteorites, asteroids, and all terrestrial planets including their satellites. For crater counting and detailed mapping, dating is often limited by difficulties such as the presence of multiply reworked surfaces, the small size of objects, or the presence of a dense atmosphere on Venus. For all dating work by nuclear methods, direct access to rock samples is required, and these samples must have a minimum size that allows mineral separation, specific chemical treatments and mass-spectrometric measurements. So far, such rock or mineral samples are available from craters on the Earth, from the Moon, and from shocked meteorites. As mentioned, the time of collision can also be deduced from the time of meteorite exposure to cosmic radiation, reflecting the time of residence in space between fragmentation of the parent body and impact of a fragment on the Earth surface. Such dating is possible on shocked and unshocked fragments.

This chapter on impact dating is focused on the determination of absolute ages, i.e., the use of nuclear methods, however, techniques and major results of geological methods (relative ages) are briefly summarized. Presentation of the different methods is followed by an overview of shock wave metamorphism in rocks and minerals, because detailed knowledge on impact produced phase transformations is fundamental for the interpretation of dating results. Well controlled sample selection in craters, distant ejecta or shocked meteorites is therefore required to successfully date an impact event. In relation to these considerations, an outlook on dating potentials is given, completed by a series of examples where dating was successful. For complementary reading and more detailed information on nuclear dating methods, the following recent review articles may be used : "*Cosmic nuclides in extraterrestrial material*" (Vogt et al. 1990), "*Dating terrestrial impact events*" (Deutsch and Schärer 1994), and "*Impact ages of meteorites : A synthesis*" (Bogard 1995).

2 Some Numbers and Definitions

4.566-4.558 Ga : Formation of the meteorite parent bodies (Allègre et al. 1995).
4.56-4.45 Ga : Accretion and differentiation of the Earth and other terrestrial planets.
4.50 Ga : Formation of the Moon.
Precambrian : Period of time between 4.56- 0.54 Ga.
Archean : Period of time between 4.56-2.70 Ga.
Proterozoic : Period of time between 2.70-0.54 Ga.
Phanerozoic : Period of time between 0.54 Ga and today.

Impactites : Rocks and minerals affected by shock wave metamorphism in relation to an impact event.
Tektites : SiO_2-rich impact melt glass, ballistic ejected over long distances from the crater area (distant ejecta).
Breccia : Rocks (or impactites) consisting of angular fragments, mostly lying in a fine-grained matrix.

Polymict breccia : Breccias containing fragments of different composition.
Autochthonous : Material that is still in its original place of formation.
Allochthonous : Material transported to the site of deposition.
Dike : A sheet-like intrusion that cuts across the host-rock.
Pseudotachylite : Amorphous, glassy rock layer or dike.
Anorthosite, Norite, Gabbro etc. : Intrusive rocks having a basalt-like composition.
Granophyre : Rock having a micrograritic structure.
Hugoniot curve (after Rankine-Hugoniot equations) : Locus of all achievable shock states of a given solid in the pressure-volume plane.

3 Geological Dating Methods

3.1 Crater Counting

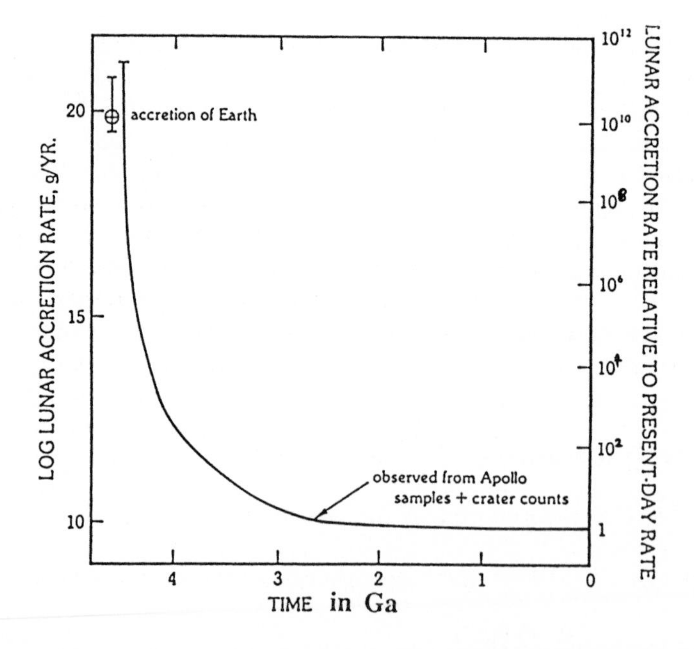

Fig. 1. Schematic rate of accretionary cratering as a function of time since primary formation of the terrestrial planets 4.56 Ga ago (simplified after : Hartman 1980; Taylor 1982]. The present day value is derived from actual observations, and the time rates are deduced from relative and radiometric dating of lunar craters.

Mass flux and size distribution of impactors in the Earth-Moon system have been established by detailed crater counts on the Moon, in combination with radiometric dating on clasts, collected by the Apollo missions in different regolith localities (for a summary see Taylor 1982). In addition, impact frequencies during Phanerozoic times (latest 0.54 Ga) were estimated by crater counting in cratonic areas such as the Canadian Shield. For the last 100 m.y., encounter probabilities could independently be estimated from astronomical recording of size and orbits of asteroids and comets (e.g. Shoemaker 1983). These observations predict the formation of 4 ± 2 craters with a diameter $D \geq 20$ km on the Earth surface every 5 m.y. To summarize such counting work, Fig. 1 schematically shows the Lunar accretion rate as a function of time since primary accretion of the terrestrial planets at 4.55 Ga (from Taylor, 1982). Note that accretion is given on a logarithmic scale illustrating the very rapid decrease by a factor of 10^{10} since the planets infancy. Concerning potential impactors, it should not be ignored that there are still more than 2000 bodies present in the Asteroid belt, in addition to comets having very different trajectories (see specific chapters in this book).

3.2 Bio-magnetostratigraphy and Paleomagnetism

So far, these techniques were restricted to the Earth as they are based on paleontological and magnetic records in sediments. In craters, paleomagnetic studies are possible, because impact formed minerals are occasionally newly magnetized along the orientation of the Earth magnetic field. With the exception of some paleomagnetic studies, application is possible only for the period of time since the appearance of extensive life forms, i.e. for the last 500 m.y. of Earth History. As pointed out, these methods yield relative ages only, unless they are combination with radiometric age dating to fully exploit their potential. Biostratigraphy, i.e. dating based on fossil records is an excellent tool to obtain initial information on the age of an crater, however, this method always dates post-impact sedimentation, and in structure where sedimentation occurs much later, the ages are necessarily too young. Such cases are craters that were first covered and then exhumed, followed by much later sedimentation in the newly exposed crater depression. High stratigraphic resolution studies of the fossil record are needed to bracket and precisely correlate ejecta layers in continental and marine environments. Such precise dating of impact phenomena in sediments is of particular importance in the evaluation of hypothesis on impact induced biological crises, and global extinction events.

4 Nuclear Dating Methods

4.1 General Aspects

Two fundamentally different approaches have to be considered : (i) dating of primary shock-related phenomena and (ii) dating of secondary features that occur in association with post-shock processes. Primary effects are phase transformations produced by shock metamorphism, brecciation and displacement of target rocks, as well as the deposition of distant ejecta. Post-impact processes comprise sedimentation in craters, hydrothermal activities, and in exceptional cases, effects to the geological environment. In general, the choice of dating methods is age dependent. This is particularly true for all techniques based on radioactive decay, because decay constants vary by up to factor of 10^8. Impact related resetting of isotopic dating systems strongly depends on (i) the amount of waste heat in a specific rock or mineral, (ii) equilibrium temperature in the impact formation that contains the shocked material, and (iii) cooling rates of the impactites.

4.2 Thermoluminescence

The use of thermoluminescence (TL) properties is a promising method to date very young, < 1 Ma old impact craters. This technique is based on the in-situ decay of radioactive elements producing lattice defects in crystals that cause characteristic variations in light emission after excitation in the laboratory. The number of defects is a function of time and radiation doses, however, it is very sensitive to crystal annealing and therefore, defects are erased with time. Shock metamorphism provides the temperatures necessary to totally reset TL in minerals having threshold conditions of about 10 GPa and 600° C. In general, sufficiently shock-heated quartz-bearing lithologies are good candidates for TL analyses but so far, TL impact dating has only been applied to sandstone from the Meteor crater in Arizona, yielding an age of 49 ± 3 ka for the impact (Sutten 1985).

4.3 Cosmogenic Nuclides

Through spallation reactions, cosmic rays produce a large number of stable and radioactive nuclides in extraterrestrial bodies, as well as in the Earth atmosphere and surface. Dating is based on the time dependent accumulation and subsequent decay of these cosmogenic isotopes, giving directly the duration of exposure to cosmic radiation or the time since later shielding. From measurements in meteorites, and results of experimental work and modeling, cosmic nuclide production rates could be determined, depending on the size, shape, and composition of the irradiated target phase. Some of the important cosmogenic nuclides are ^{3}He and ^{21}Ne (both stable), ^{10}Be, ^{26}Al, ^{36}Cl (Half-lives 0.3-1.6 m.y.), and ^{22}Na and ^{60}Co (Half-lives < 10 years). Due to such a large spectrum of different nuclide stabilities, parent-body fragmentation can be dated over a large range of ages. For chondrites, exposure ages range from ≈ 0.1 to 100 Ma, and for iron meteorites they are significantly older lying between ≈ 10 and 1000 Ma. Such longer exposure of iron meteorites is probably an effect of their stronger mechanical resistance to fracturing, i.e. a smaller number of new surfaces are produced during a collision event in space. As mentioned, if the meteorite has a simple cosmic ray exposure history, these ages correspond to the period of time between the collision (excavation) of the meteorite parent-body, and its impact on Earth. This simple picture does not apply if several collisions have changed cosmic production rates by changing the shape, size or shielding of the exposed fragment. In detail, production rates depend on the concentration of the target element, reaction specific cross sections, as well as on energy and flux of the cosmic ray particles. For a review of physical parameters and application of cosmogenic nuclides to determine exposure ages, more information can be found in Vogt et al. (1990).

The radiation exposure principle is also of interest for dating impact events on Earth, however, its application is restricted to craters less than a few Ma old, because only nuclides with short half-lives are produced on continental surfaces (^{10}Be, ^{14}C, ^{26}Al, and ^{36}Cl). The different methods available are all based on the accumulation of cosmogenic nuclides in excavated material, which was shielded against cosmic rays prior to ejection. In rare cases, relicts of the impactor have been preserved in Prehistoric, relatively small ($D < 1$ km) impact craters allowing direct dating of the impact event by measuring the terrestrial exposure age of the meteorite fragments. The technique was successful for the Arizona crater giving concordant ^{36}Cl and ^{10}Be–^{26}Al ages of 50 ± 1 ka (Nishiizumi et al. 1991; Phillips et al. 1991) that are in agreement with the TL age of 47 ± 3 ka from the same site. Recent developments in surface exposure dating may extend this technique to other cases.

4.4 Fission Track Dating

Counting of accumulated tracks, induced by spontaneous fission of ^{238}U in minerals and glasses is a widely used tool to investigate the cooling of rocks, and it has gained importance for impact dating. Such tracks are stable below mineral-specific annealing temperatures, which are about 80° C for apatite and 240-260° C for zircon and titanite (Wagner and Van den haute 1992). In its application to impact dating, this method yields only correct ages if all pre-existing tracks were completely erased by the impact event. Another limiting factors in track dating are low U-contents, because ^{238}U fission probability is only about 10^{-17} y^{-1}, and relatively high concentrations of U in minerals and glasses are required (> 10 ppm). For old craters where deep levels are exposed, partial fading out of tracks is caused by normal thermal gradients prior to denudation, producing too young ages. About 20 impact craters on Earth have been dated so far by track counting, and the ages cover a time-range from 4 ka for the Henbury craters, Australia, up to 300 Ma for the Clearwater Lake crater, Canada (e.g. Wagner and Van den haute 1992). The most commonly used material is impact melt glass in craters, although this material is known to be very sensitive to track fading. Tektites are also excellent material for fission track dating, implying the same type of corrections for fading. In addition to these glass lithologies, minerals separates from shocked but unmelted rocks can also be used. Examples of this application are given by apatite dating in the central uplift of the about 65 Ma old Manson structure in Iowa (Hartung et al. 1986), and from glass in a shocked gneiss fragment from the polymict breccia in the 23 Ma old Haughton structure, Canada (Omar et al., 1987).

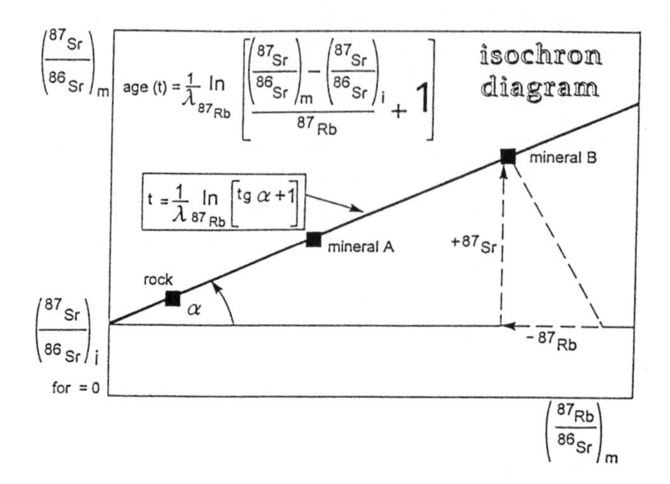

Fig. 2. Example of a Rb-Sr age determination using the isochron diagram. The age is defined by the slope of the regression line through the individual data points, which must be measured on cogenetic phases having different parent/daughter element ratios. The atomic abundance of the radioactive (^{87}Rb) and the radiogenic isotope produced (^{87}Sr) have to be expressed relative to a stable isotope of the daughter element (^{86}Sr), and the Y-axis intercept of the isochron gives the initial ^{87}Sr/^{86}Sr composition of Sr, incorporated into the cogenetic phases at the time of mineral growth.

4.5 Radioactive Decay

As for conventional geochronological and cosmochronological dating purposes, the following decay systems are potentially available for impact dating :

Isotope system	*Age diagram to apply*
$^{87}\mathrm{Rb} \Longrightarrow {}^{87}\mathrm{Sr}$	isochron
$^{40}\mathrm{K} \Longrightarrow {}^{40}\mathrm{Ar}$ (^{39}Ar–^{40}Ar)	plateau-age and isochron
$^{147}\mathrm{Sm} \Longrightarrow {}^{143}\mathrm{Nd}$	isochron
$^{238}\mathrm{U} \Longrightarrow {}^{206}\mathrm{Pb}$	isochron or concordia (combination with ^{235}U–^{207}Pb)
$^{235}\mathrm{U} \Longrightarrow {}^{207}\mathrm{Pb}$	isochron or concordia (combination with ^{238}U–^{206}Pb)
$^{232}\mathrm{Th} \Longrightarrow {}^{208}\mathrm{Pb}$	isochron
$^{176}\mathrm{Lu} \Longrightarrow {}^{176}\mathrm{Hf}$	isochron
$^{187}\mathrm{Re} \Longrightarrow {}^{187}\mathrm{Os}$	isochron

So far, only the first five clocks were used for impact dating but all of them have Half-lives that lie in the order of the age of the solar system and in consequence, they may be used to date impact events between 4.55 Ga and less than 1 Ma. In general, the age of the impact can only be obtained if parent and daughter isotopes (elements) are re-homogenized during the impact event, i.e. the radiometric system is totally reset to a zero age. An exception is the U-Pb chronometer which includes two decay systems, allowing the determination of ages also for cases where the system was not totally reset or re-opened by post-impact processes.

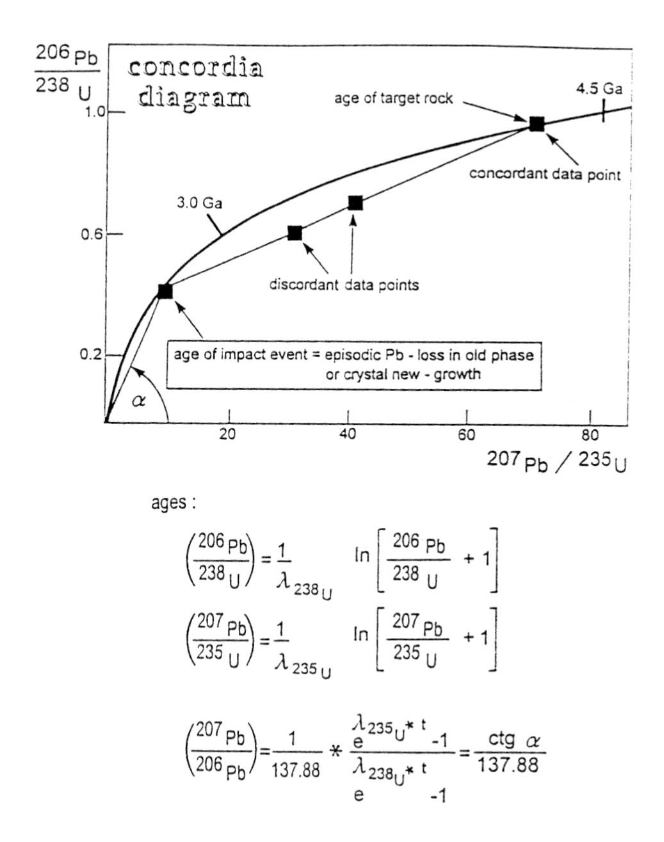

$$\left(\frac{^{206}Pb}{^{238}U}\right) = \frac{1}{\lambda_{^{238}U}} \ln\left[\frac{^{206}Pb}{^{238}U} + 1\right]$$

$$\left(\frac{^{207}Pb}{^{235}U}\right) = \frac{1}{\lambda_{^{235}U}} \ln\left[\frac{^{207}Pb}{^{235}U} + 1\right]$$

$$\left(\frac{^{207}Pb}{^{206}Pb}\right) = \frac{1}{137.88} * \frac{e^{\lambda_{^{235}U} * t} - 1}{e^{\lambda_{^{238}U} * t} - 1} = \frac{\mathrm{ctg}\ \alpha}{137.88}$$

Fig. 3. Concordia diagram with the $^{207}Pb/^{235}U$ and $^{206}Pb/^{238}U$ ratios on the Y and X axis, respectively. In this diagram, ^{207}Pb and ^{206}Pb reflect the accumulated radiogenic components, corrected for initial common Pb. The age is given either by concordant data or by concordia intercepts of regression lines through discordant data. This combination of the two U-Pb chronometers is the only method available to investigate open system behavior, i.e. to obtain ages for only partially reset, only disturbed systems.

To illustrate these dating methods, Fig. 2 shows a schematic isochron diagram for the Rb-Sr system, and Fig. 3 gives the concordia diagram, which can only be used for the for U-Pb chronometer. It has to be emphasized that most ages are obtained through these two diagrams, with the exception of ^{40}Ar-^{39}Ar, where ages can additionally be derived from stepwise outgassing of neutron irradiated samples (to produce ^{39}Ar from ^{39}K), yielding plateau ages such as illustrated in Fig. 4a, for a biotite from an ordinary granite. This plateau ages can also be expressed in the form of a normal, equivalent isochron plot such as shown for the very same biotite in Fig. 4b. As evidenced by the isochron and plateau age presentation, both the age and initial isotopic composition (of the daughter element) have to be obtained through graphical treatment, where the age is given by the slope of the isochron, and the initial isotopic compositions by its intercept with the Y-axis.

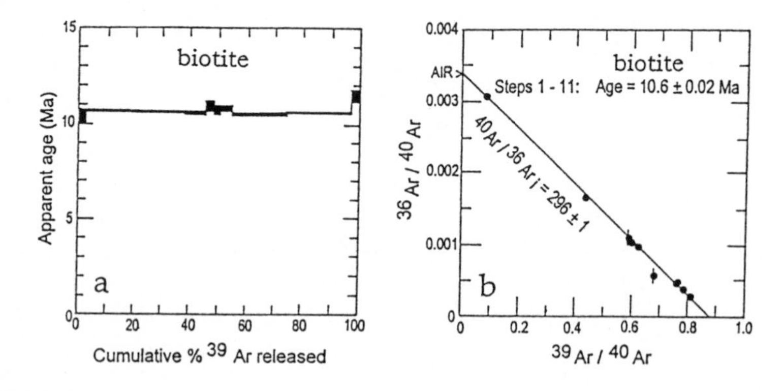

Fig. 4. Example of a plateau age (a) and its corresponding isochron plot (b) such as measured by the ^{40}Ar-^{39}Ar method. The plateau is obtained by stepwise outgassing of the irradiated sample, with the different steps corresponding to the individual data points in the isochron diagram. The example given is a biotite from a young granite (from : Schärer et al. 1990). Ages are given on the Y-axis for the plateau presentation, and by the slope of the regression line in the isochron diagram, the intercept of which with the Y-axis defines the initial isotopic composition of argon in the sample.

To obtain reliable ages in the isochron diagram, all phases analyzed (e.g. the rock and different minerals) must have formed coevally, and intitial isotopic compositions must have been identical in all phases. This means that all phases were simultaneously extracted from a common, and isotopically homogeneous reservoir. Since many impact shocked rocks are heterogeneous systems, these boundary conditions are often not fulfilled, and the age (impact) significance of an isochron age remains ambiguous. This is particularly true for many shocked meteorites, for which apparent age patterns cannot be ascribed to resetting during a distinct shock event. From a mineralogical point of view, such heterogeneity of impactites is often reflected by complex mineralogical composition and inter-growth relationships, precluding adequate separation of pre-shock and newly formed, equilibrated phases (e.g. Schärer and Deutsch 1990; Martinez et al. 1991).

In contrast to the isochron diagram, to derive ages in the U-Pb concordia diagram, all data have to be corrected for initial Pb prior to plotting, and the ages are given either by data that plot on the concordia curve (concordant data) or by intercepts of regression lines through discordant data. These cases are illustrated in detail in Fig. 3. To correct for initial lead, isotopic analyses on cogenetic minerals devoid of uranium are needed, or model isotopic compositions for distinct source rocks have to be calculated for varying model parameters.

5 Shock Wave Metamorphism

5.1 General Aspect

During impact, most of the impactors kinetic energy is transferred to the target rocks, where it is transformed to internal and kinetic energy. These energies cause very high pressures and temperatures, as well as ejection of shocked material over large distances from the impact site. Such a scenario is schematically illustrated in Fig. 5, distinguishing the different stages that characterize an impact event. In addition, Fig. 6 summarizes the pressure-temperature conditions reached during shock wave metamorphism, also indicating phase transformations that are indicative for distinct shock conditions. Such transformations are irreversible above the Hugoniot elastic limit, lying in the range of 2-12 GPa for silicates. Shocked minerals are characterized by progressive breakdown of crystallographic orders and by non-equilibrium structural states. Classification of shock stages (0-V : Stöffler 1971) is based on these characteristic shock effects, which occur, for constant pressure, at much higher temperatures in porous minerals than in non porous phases. For example, at a shock pressure of about 10 GPa, dense silicates are fractured at roughly 100° C only, whereas porous silicates are transformed to diaplectic glass at 400° C. Shock conditions exceed by orders of magnitude pressures and temperatures reached during endogeneous processes such as magmatism and metamorphism of crustal rocks. At pressure above 200 GPa, temperatures of several 1000° C are reached, and target material is totally volatilized (shock stage V).

To understand the occurrence of the different types of impactites (impact formations), i.e. the location of characteristic shock products in a crater, Fig. 7 shows a typical crater such as to expect on all terrestrial planets (e.g. Melosh 1989). Differences in this schema will concern the distance of ejecta layers due to variations of gravity and density of the atmosphere. Other differences may be expected for chemical reactions of the very hot impactites with atmospheric components, including water. For impact dating, such differences are of minor importance, possibly with the exception of the K-Ar system on Earth, where impact related implantation of atmospheric Ar may cause major problems. The commonly distinguished impact formations are :

(1) coherent impact melt layers
(2) allochthonous breccia deposits
(3) crater basement
(4) distant ejecta

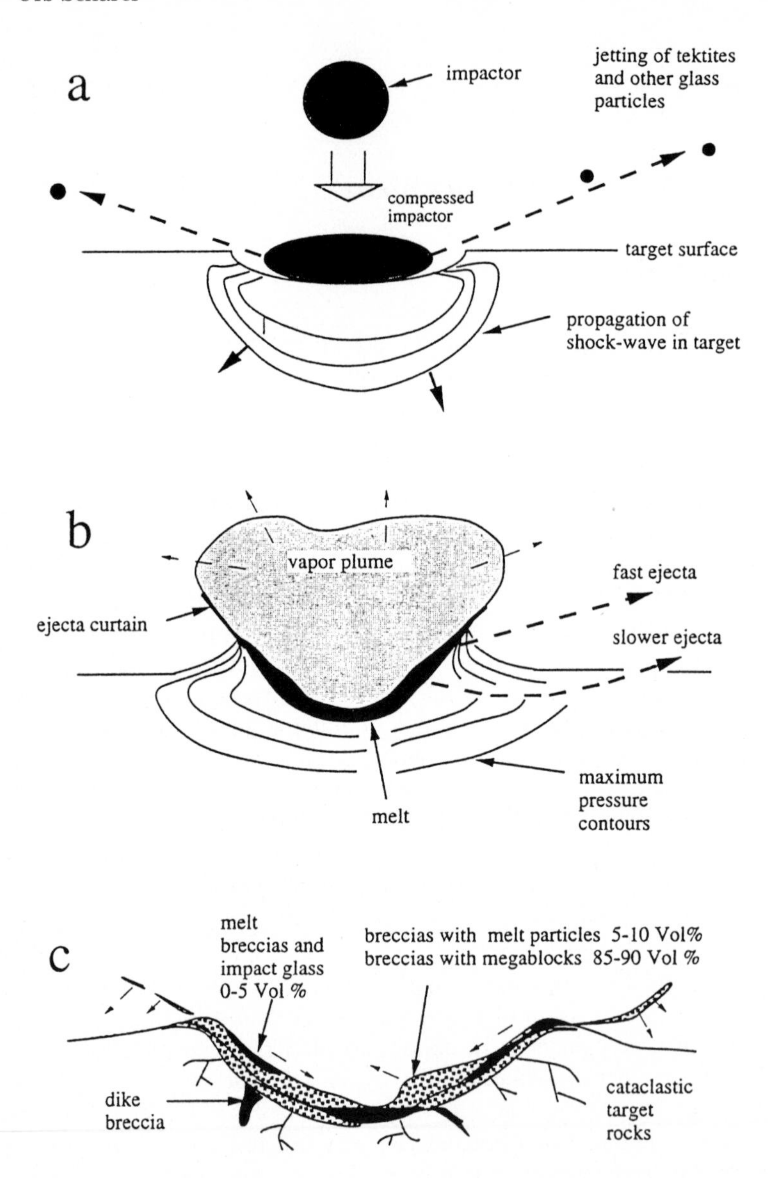

Fig. 5. Schematic scenario of an impact producing a transient crater of large size (> 10 km, modified after : Melosh 1989), showing (a) initial contact of the projectile and the jetting of target surface melt (tektites), (b) formation of the vapor plume, associated with the production of fast ejecta and shock metamorphism in the target rocks, and (c) the formation of a transient crater containing melts and different types of breccias to various proportions.

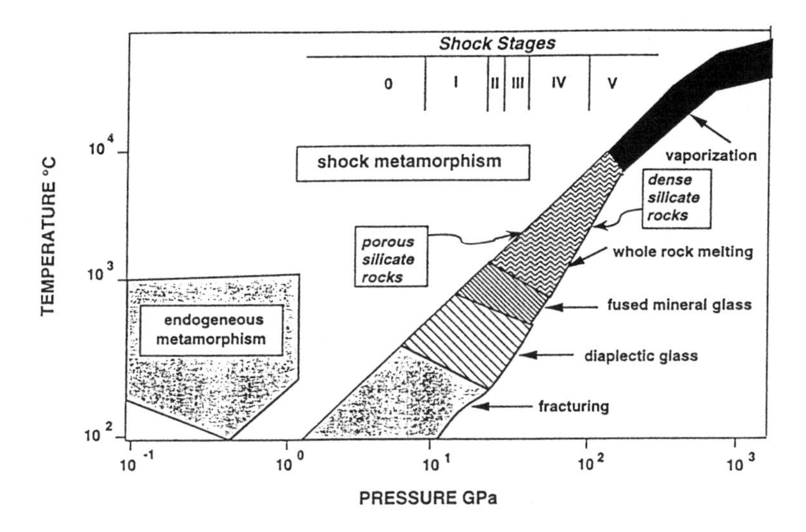

Fig. 6. Pressure-temperature range of shock metamorphism up to total volatilization of target material. The upper and lower limits of the P-T field for shock metamorphism are defined by porous and dense silicates, respectively, reaching much higher shock temperatures in porous material for a given shock pressure (from : Martinez et al. 1993; Stöffler 1971). The figure also indicates phase transformations that are typical for distinct shock stages, e.g. the formation of diaplectic glass (solid state amorphization), fused glass (shock melting), and vaporization of the rock.

5.2 Coherent Impact Melt Layers

They are usually composed of clast-free impact melt rocks and polymict melt breccias, containing shocked and unshocked rock fragments in varying proportions. The fragments are embedded in a glassy or re-crystallized melt matrix, occasionally showing coarse grained "magmatic" textures. Melting on the whole rock scale is induced by high post-shock temperatures at shock pressures ≥ 60 GPa (e.g. Stöffler 1971; Grieve et al. 1977). Good examples for coherent impact melt sheets exist in the 100 km large Manicouagan structure (Canada), where up to 230 m of the melt layer is preserved, and at Sudbury (Canada), where the Sudbury "Igneous" Complex represents an even larger, differentiated impact melt system. The time scale of cooling is dependent on fragment content and thickness of the melt layer, and cooling rates are variable, with the central part of the melt sheet remaining hot much longer than the clast-rich marginal facies.

5.3 The Crater Basement

It is usually composed of monomict and brecciated mega-blocks and parautochthonous breccias, both injected by dikes. These dikes represent impact melt, pseudotachylites or fragmental breccias. Maximum shock pressures recorded in the crater basement are about 45 GPa. The post-impact thermal history in the crater basement is determined by (1) post-shock waste heat of the shocked

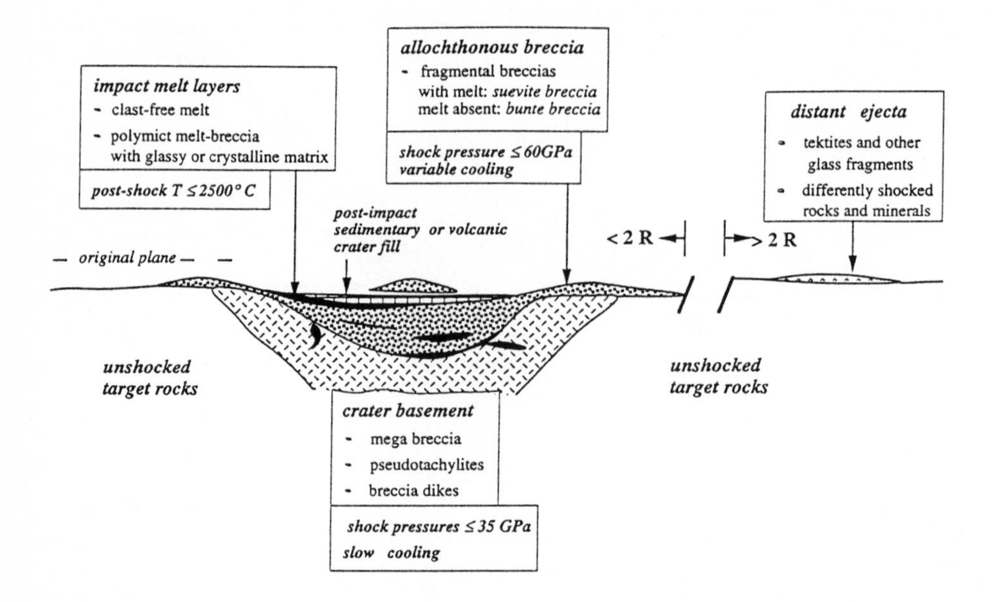

Fig. 7. Schematic location of distinct impact produced rocks and melts, also indicating some P-T shock metamorphic conditions that are typical for these impactites. The subunits distinguished within each formation are commonly used terms for impact shocked rocks (adapted from : Melosh 1989; Deutsch and Schärer 1994). Note that distant ejecta from large craters (> 20 km) may occur at several 100 km distance from the crater rim.

material, and (2) size of the overlying crater fill, which may comprise impact melt. In addition, voluminous melt sheets cause a reverse temperature gradient in the crater floor. In cases, where the transient cavity reaches deep crustal levels, cooling rates in the basement also depend on the ambient geothermal situation, and temperature decay is slow compared to all other impact formations. Such slow cooling is potentially an advantage for dating, because diffusion kinetics may remain sufficiently rapid during a long period of time, allowing the isotope systems to be reset even in not melted but highly shocked target rocks.

5.4 Allochthonous Breccias Deposits

This impact formation is characterized by breccias occurring either within the crater or up to two crater radii away. Mineral and rock fragments in the breccias display a wide range of shock metamorphic features and melt particles may be present. Therefore, allochthonous breccias in a given site have usually experienced very varying post-shock temperatures and thermal equilibrium is reached quickly after deposition. Cooling heavily depends of the abundance of melt particles, and equilibrium temperatures may exceed 600° C. Observations on various breccias of the Ries crater (Germany) indicate that cooling to ambient conditions takes place on the order of months to a few hundred years (e.g. Staudacher et al.

1982). More complex thermal histories in allochthonous breccias may be caused by post-shock processes such as circulations of hot gas and fluids, ascending from deeper levels of the crater or the basement.

5.5 Distant Ejecta

Distant ejecta occurs up to hundreds or thousands of kilometers away from the crater, having suffered rapid quenching during ejection. The K/T boundary clay represents a well documented example for this type of impact formation, which also comprises variably shocked rocks and minerals, impact melt glass, tektites and microkrystites, which may be high temperature condensates from the ejecta vapor plume.

6 Implications from Experiments

Shock experimental results show that the post-shock thermal history is the major controlling factor for the behavior of isotopic systematics in shocked rocks. In consequence, any dating approach strongly depends on the impact formations available in a specific impact site or in shocked meteorites. If shock metamorphism has not totally re-equilibrated the phases, dating results may give either (i) pre-shock ages, (ii) the impact age, (iii) the age of post-impact processes or (iiii) no age information, because shock and post-shock history caused disturbances of the chronometer only. Shock experiments have allowed to define major boundary conditions to successfully date impact events, although shock wave passage is much shorter in the experiments, and cooling from post shock temperatures is very fast. For a summary and more detailed presentation of shock recovery experiments we refer to earlier publications (e.g. Deutsch and Schärer 1990; 1994). In none of these experiments a total resetting of the isotope clock could be observed, even in shocks reaching melting conditions. In consequence, total rock melting is required to produce rock-scale equilibrium conditions in impact formations. On the other hand, long lasting high temperature regimes in impactites may allow to equilibrate systems that were not in equilibrium immediately after shock-wave passage. Such annealing, associated with high subsolidus diffusion rates possibly reset Rb-Sr and K-Ar systems on the cm-scale.

7 Dating and Sample Selection

7.1 Geological Frame

To illustrate dating potentials in shock metamorphosed rocks, Fig. 8 attaches possible dating methods to characteristic impact lithologies of large craters (> 10 km). For any sample selection in and around the crater it must be considered that more than 90 % of the shocked rocks have preserves its pre-shock age, because shock and post-shock conditions are not sufficient to reset the chronometers. Within the remaining 10 % of impactites, the isotope systems show varying

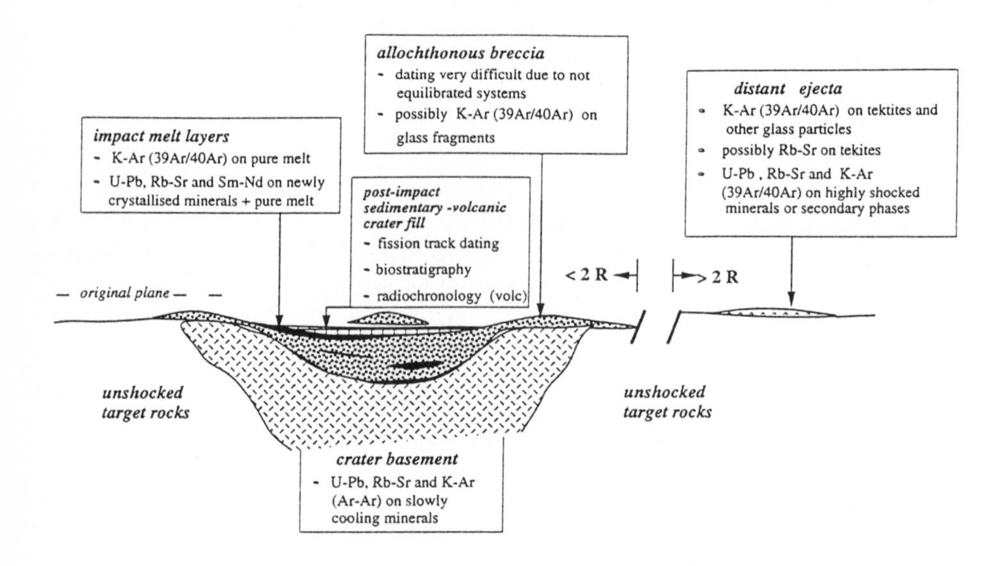

Fig. 8. Potentially successful dating approaches for different impact formations in large craters, including later crater fill such as sediments (Earth) and volcanic layers (all terrestrial planets and the Moon).

effects, and only a very small portion is susceptible to yield the correct impact age. These boundary conditions reduce successful sample selection to less than 1 target material. In certain cases, the age may be obtained by dating minerals that formed during secondary processes that affecting the crater or distant ejecta. Potentially datable minerals could form in hydrothermally active zones or in crater lakes. If secondary processes occur within 0.1 m.y. after the impact, the age obtained still reflects the event, because accuracy on radiometric ages is rarely better than 0.1 Ma. As mentioned earlier, first age constraints for young terrestrial craters are often given by fossils that populate crater depressions or sediments contained in distant ejecta.

Given the knowledge from experimental work and earlier dating studies, all melt lithologies are good candidates for radiometric dating. Such melts are not restricted to the coherent melt sheet, and melt inclusions in suevitic breccias, and tektites are equivalent samples. In general, melting on the total rock scale produces homogeneous melt domains from which equilibrated glass, minerals and small rock samples solidify, having entirely reset radiometric clocks. The same conditions also apply for meteorite samples, with the limiting difference that such samples are in general small compared to terrestrial craters, and minerals and glass have to be extracted from a very small amounts of shocked material (milligrams). The actual scale of isotopically homogeneous domains may vary by orders of magnitude, and specific dating possibilities have to be evaluated for each case.

As emphasized, annealed samples from other melt-containing impact formations are also available for dating. To produce such cases, the highly shocked material must have cooled slowly from high equilibrium temperatures to allow solid state diffusion to equilibrate the phases. This constraint even applies for rocks shocked at pressures near the onset of whole rock melting. Candidates for this type of sample selection are footwall rocks lying directly beneath the coherent impact melt layer. Moreover, equilibrated systems may be found in the slowly cooling crater basement, which remained covered for a long period of time after the event. In considering all these boundary conditions, the following dating possibilities can be proposed :

7.2 Coherent Impact Melt Layers

(1) U-Pb dating on newly crystallized accessory phases such as zircon ($ZrSiO_4$), baddeleyite (ZrO_2), titanite (Ca,Ti-silicate) and rutile (TiO_2) from the impact melt. In some cases, inherited, not totally melted components may limit the precision of dating.
(2) Rb-Sr dating on minerals that formed newly in the melt, without any incorporation of old components. Such dating requires a grain by grain selection from a given melt domain, accompanied by chemical and crystallographic test analyses. In certain cases, analyses on small glassy domains may be included. In voluminous impact melt sheets, dating with Sm-Nd mineral isochrons may also be successful if adequate minerals separates can be selected from the rock.
(3) Fission Track and K-Ar (^{40}Ar-^{39}Ar) dating of clast-free melt rocks may yield excellent results. Since K-Ar and Fission Track ages record closure temperatures that are significantly lower than post shock temperatures, they yield information on the cooling history of a melt rock. In cases of fast cooling, all geochronometers should give the same ages, dating the impact event with high precision.

In all these dating attempts, problems arise from the presence of very small, isotopically not equilibrated clasts that cannot be distinguished and separated quantitatively. Some caution also concerns K-Ar and Fission Track data, because both systems are very sensitive to alteration at low temperatures and therefore, secondary processes may cause too young ages.

7.3 Allochthonous Breccia Deposits

In this lithology, differently shocked target rocks may be welded together with shock-produced glass, reaching rapidly thermal equilibrium at relatively high temperatures. Due to the co-presence of shocked, unshocked, and/or glassy components to various proportions, such ejecta deposits have complex temperature histories, and sample selection is difficult. Nevertheless, melt particles in breccias are promising objects for Fission Track and ^{40}Ar-^{39}Ar dating. Slow cooling from high formation temperatures may cause a re-equilibration of the K-Ar decay system and a total annealing of pre-existing fission tracks. Under these conditions, even weakly shocked material can yield reliable impact ages, in contrast to highly shocked but quickly cooled lithologies. The cosmogenic nuclide method

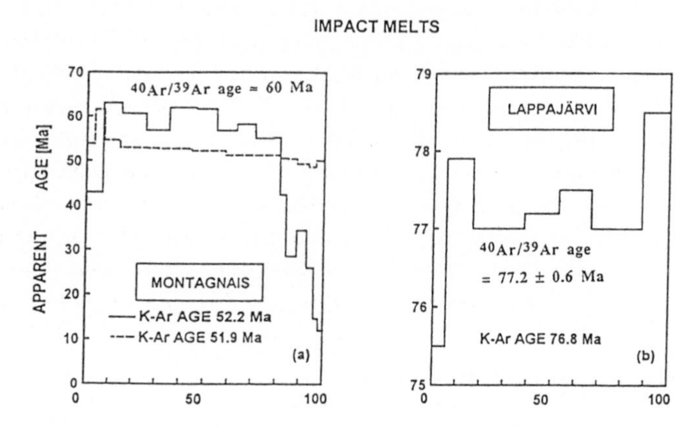

Fig. 9. ^{40}Ar-^{39}Ar outgassing spectra for fine-grained impact melt rocks (a) from the about 60 Ma old submarine Montagnais structure (after : Bottomley and York 1988), and (b) for the 77 Ma old Lappajärvi crater (Jessberger and Reimold 1980). Both spectra do not define "prefect" plateau ages, illustrating the difficulty to obtain precise ages even on impact melts. Such difficulties may be due to the implantation of atmospheric Ar, irradiation problems with very fine-grained material, or other reasons such as the presence of sub-microscopical relicts from the target rocks.

is potentially feasible for the dating of ejecta deposits in young craters. This approach is unrelated to the post-shock thermal history of the ejected material but it requires particularly delicate sample collection relative to real surface exposure, rock orientation, and erosion rate.

7.4 Crater Basement

So far, rocks that constitute the crater basement have rarely been used for impact dating, however, examples from the large impact structures Manicouagan and Sudbury are encouraging. In polymict breccias, internal Rb-Sr and Sm-Nd mineral isochrons and U-Pb analyses on newly grown accessory minerals may yield the impact age. An important condition for the resetting of isotopic clocks in the crater basement is the presence of a thick impact melt layer causing thermal metamorphism and slow cooling of the underlying crater floor lithologies. Fission Track and ^{40}Ar-^{39}Ar dating may be also be successful, because both chronometers can be reset at relatively low temperatures. Pseudotachylites in the crater basement are good candidates for this approach, and in deeply eroded structures, they are the only candidates for impact dating. Due to the fine-grained or glassy nature of the matrix, pseudotachylites easily re-crystallize at low temperatures causing re-opening or resetting of isotopic systems.

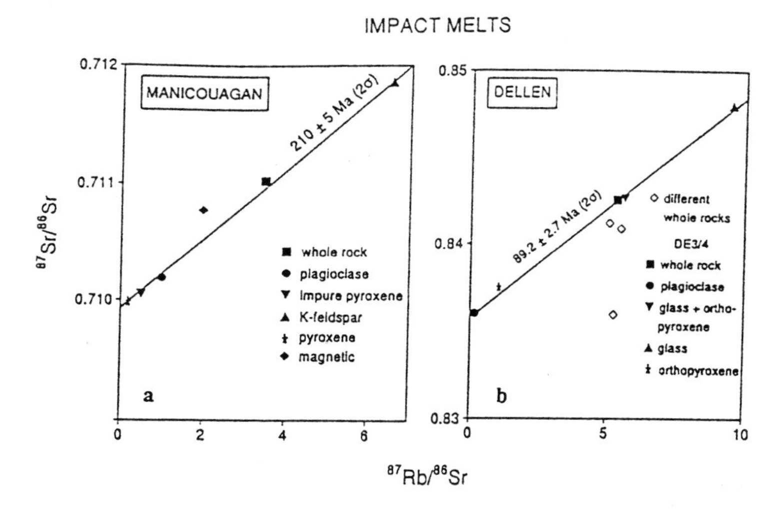

Fig. 10. Two examples of internal Rb-Sr isochrons from the about 210 Ma old Manicouagan (a), and the about 89 Ma old Dellen (b) impact structures (Jahn et al. 1987; Deutsch et al. 1992, respectively). To obtain such isochrons, the different minerals must be selected grain-by-grain from the impact melt, in avoiding any contamination from relictic, unmelted target material.

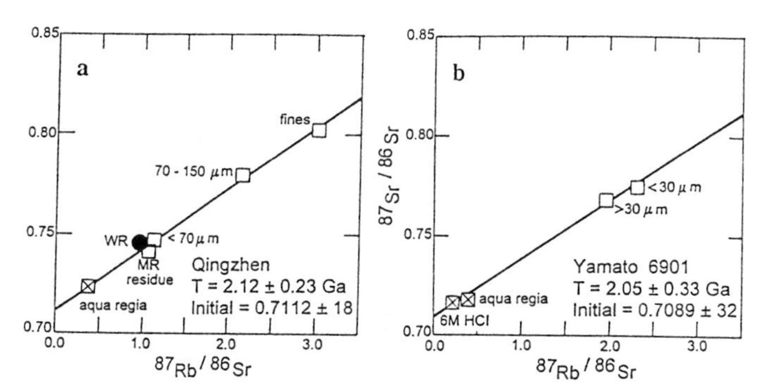

Fig. 11. Rb-Sr age diagrams (simplified after : Torigoye and Shima 1993) of two meteorites for which 30-150 mm size particles and acid-leach fractions define internal isochrons. These isochrons can be interpreted as small-scale re-equilibration of the chronometer during a thermal event that affected the rocks 2.4 Ga after primary parent body formation. In consequence, the ages of 2.12 ± 0.23 and 2.05 ± 0.33 most likely reflect a impact event that occurred on the parent bodies of the Qingzhen and Khairpur chondrites.

7.5 Distant Ejecta Deposits

Tektites are near-ideal objects for Fission Track and 40Ar-39Ar impact dating, because their very high formation temperatures efficiently melt and re-homogenize potentially heterogeneous melt domains. Such high temperatures also cause quantitative Ar-loss and annealing of fission tracks. In some rare cases, inherited Ar components or fragments of the target rock may yield too old ages.

8 Examples

8.1 Impactites of the Crater Area

For terrestrial craters, only whole rocks samples were used so far for ^{40}Ar-^{39}Ar dating of impact melt rocks. This may be the reason for the small number of well-defined Ar ages. The technique was relatively successful for the Montagnais and Lappajärvi impact structures (Bottomley and York 1988; Jessberger and Reimold 1980) yielding ages of about 60 and 77.2 $\pm$ 0.6 (2σ) Ma for the two craters, respectively. The ^{40}Ar-^{39}Ar degassing pattern for these two examples is displayed in Fig. 9, which also summarizes K-Ar ages obtained on the same rocks. For the Montagnais case, K-Ar ages are by about 10 Ma younger, whereas they are identical for the melt rocks from the Lappajärvi crater.

Concerning Rb-Sr dating on such melts, examples of reliable impact ages were reported for the Manicouagan and Dellen impact structures (Jahn et al. 1978; Deutsch et al. 1992). These two cases of internal Rb-Sr isochrons are given in Fig. 10. Problems with this approach in melt rocks may arise from enhanced alkali element migration and changes induced by later hydrothermal processes. An example of such alteration is reported from the upper part of the melt system at Sudbury, where the granophyric lithology of the Sudbury "Igneous" Complex shows geologically unrealistic initial $^{87}Sr/^{86}Sr$ values, not yielding any reliable impact age (Gibbins and McNutt 1975; Deutsch 1993).

An example for Rb-Sr systematics in extraterrestrial material is given in Fig. 11. This figure shows data obtained for two enstatite chondrite meteorites, for which 30-150 μm size particles and their acid-leach fractions define internal isochrons (Torigoye and Shima 1993). The existence of these isochrons can be ascribed to small-scale re-equilibration of the chronometer during a thermal event affecting the rocks at about 2.1 Ga. This event may be interpreted as shock-heating by an impact, occurring on the meteorite parent bodies. Not shown in the figure are a few leach fraction and rock analyses, which did not plot on the isochron. Such heterogeneities are due to the fact that re-equilibration on the rock-scale is only reached in totally melted rocks (e.g. Deutsch and Schärer 1994). In the cases of the Qingzhen and Khairpur meteorites these conditions were not reached, and re-equilibration occurred in very small rock domains only, leaving behind a essentially unequilibrated meteorite.

In voluminous impact melt sheets such as present in the Sudbury structure, U-rich minerals such as zircon and baddeleyite newly crystallize from the melt, yielding concordant or nearly concordant ages in the concordia diagram. The example for the Sudbury melt sheet is given in Fig. 12, showing a series of zircon and baddeleyite analyses that define a precise age of 1580 $\pm$ 1 Ma for the impact event (Krogh et al. 1984). Dating by accessory phases requires that most of the zircons and other U-rich minerals were totally resorbed or melted in the target material, and careful grain-by-grain selection is needed to separate newly

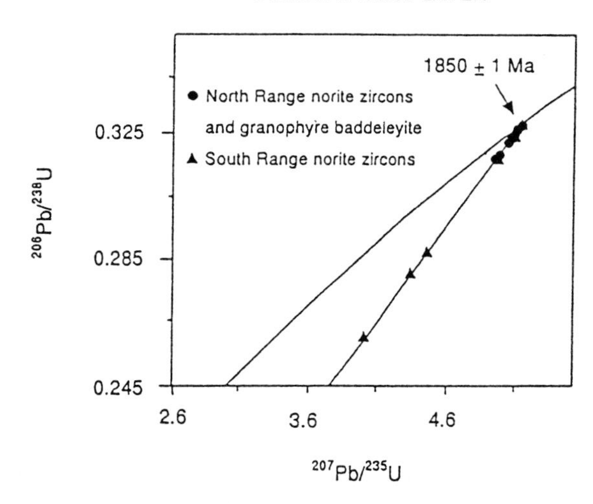

Fig. 12. ^{40}Ar-^{39}Ar ages (Maurer et al. 1978) of four selected basaltic and anorthositic 2-4 mm large breccia fragments collected at the Apollo 16 landing site. The basaltic fragment (a) yields the youngest age of 3.76 ± 0.04 Ga, whereas the three anorthositic samples define ages of 3.90 ± 0.04 (b), 3.96 ± 0.04 Ga (c), and 4.00 ± 0.04 Ga (d). All these ages date large, basin-forming impact events on the Moon producing craters that are at least a few 100 km in diameter.

grown zircons from possible relicts of older grains. Any relict zircon present in the newly formed grains would produce too old ages (inherited radiogenic lead).

8.2 Distant Ejecta, and the Cretaceous/Tertiary (K/T) Boundary

To date tektites, the K-Ar method was extensively used since 1950 (e.g. Suess et al. 1951). Occasionally such attempts are complicated by the presence of large amounts of atmospheric argon, occurring either in vesicles or in the glass mass itself (e.g. Jessberger 1982). In some glasses, such trapped components seriously hampers dating by the K-Ar or ^{40}Ar-^{39}Ar methods, however, in most cases, incremental heating or laser step heating of tektites or glass spheruls define acceptable and precise plateau ages. A good example for this dating method are the North American tektites, for which an age of 34.7 ± 0.4 Ma was determined (Bottomley and York 1991). These tektites were probably produced by at least two impacts, however, given the analytical precision they must have occurred within an interval of less than 0.4 m.y. Although these micro- and macro-tektites are abundant and widespread over the continent and the Pacific ocean floor, their source crater are still unknown.

Pioneering ^{40}K/^{39}Ar work on lunar melt rocks from the Apollo 14 site allowed to estimate the age of the very large Imbrium impact event. Impact fragments and melts were sampled at a distance of about 550 km from the crater rim, and the age obtained on KREEP-rich melts is 3.82 Ga, which is agreement

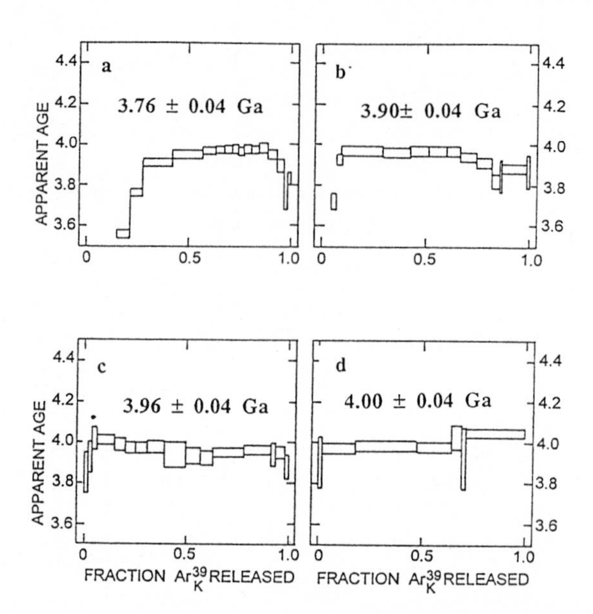

Fig. 13. Concordia diagram for U-Pb dating results for zircon ($ZrSiO_4$) and baddeleyite (ZrO_2) from the Sudbury impact structure, yielding an age of 1.85 Ga for the impact event. These two minerals have newly crystallized from the Sudbury impact melt, which was produced by melting of about 2.7 Ga old target rocks. The 1.85 Ga age is defined by the upper intercept of variously discordant mineral fraction that have suffered different degrees of Pb-loss in recent times (after : Krogh et al. 1984).

with ages obtained on basaltic clasts from the same site (for a summary, see Taylor 1982). Further dating on breccia fragments from the Apollo 16 landing site allowed to establish a multi-impact lunar cataclysm (Maurer et al. 1978) including formation of the Nectaris crater at 3.97 Ga, followed by a series of basin-forming events within a relatively short period of time ($\approx$ 0.1 Ga). A few example of this important $^{40}K/^{39}Ar$ work are given in Fig. 13 showing four selected degassing spectra of basaltic and anorthositic 2-4 mm large breccia fragments. The basaltic fragment (Fig. 13a) yields the youngest age of 3.76 $\pm$ 0.04 Ga, whereas the three anorthositic samples define ages of 3.90 $\pm$ 0.04, 3.96 $\pm$ 0.04 Ga, and 4.00 $\pm$ 0.04 Ga (Fig. 13 b, c and d, respectively). The occurrence of impact fragments of different age excellently illustrates the multi-episodic impact origin of the lunar soil at the Apollo 16 site, and exposure-age dating on the same samples (Maurer et al. 1978) shows that the last breccia-forming event occurred only recently, at $\approx$ 50 Ma.

As mentioned, accuracy and precision requirements of impact dating became strikingly clear in the context of discussions concerning impact related mass extinction at the end of the Cretaceous period. Any hypervelocity collision susceptible to trigger such extinction must have occurred during a short period of reverse polarity in the Earth magnetic field, constrained between 65.4 and 64.7 Ma (LaBrecque et al. 1977). The time of this magnetic reversal is in good

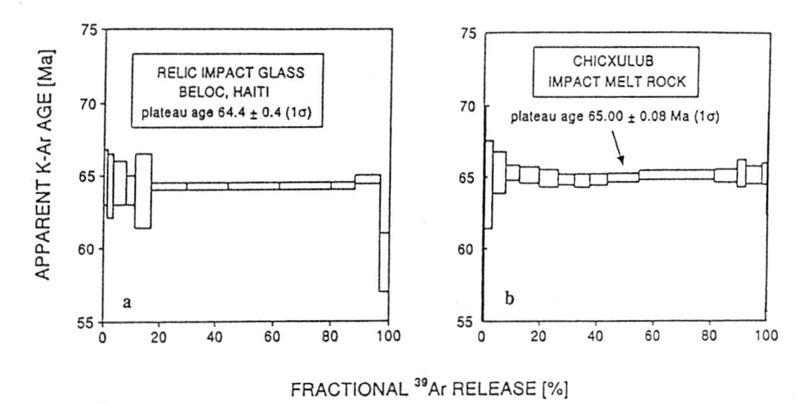

Fig. 14. (a) : 40Ar-39Ar ages for glass from the distant ejecta in the Beloc formation (Haiti), and (b) the identical age for impact melt glass from the likely Chicxulub source crater in Yucatan, Mexico (modified from : Izett et al. 1991; Swisher et al. 1992). Note that both ages are within analytical uncertainty identical with the age of the K/T boundary.

agreement with the 64.4 ± 1.2 Ma age obtained for a bentonite horizon that lies less than 1 m above the K/T boundary clay (Baadsgaard et al. 1988). Due to the presence of impact produced glass in the boundary layer itself, the dating test was additionally successful yielding precise ^{40}Ar-^{39}Ar ages between 64.5 ± 0.2 and 64.9 ± 0.1 Ma for the Beloc (Haiti) glasses, and 65.1 ± 0.2 Ma for the Arroyo el Mimbral (Mexico) case (Izett et al. 1991; Swisher et al. 1992; Hall et al. 1992; McWilliams et al. 1992). These ages clearly establish that the impact event occurred within the required period of time, and it definitely marks the K/T boundary. Fig. 14a shows an example for one of the Beloc impact glass samples , and Fig. 14b gives the equivalent age obtained for melt rocks from the suspected Chicxulub source crater in Mexico (Swisher et al. 1992).

Complementary constraints for the K/T boundary impact event was reported by Krogh et al. (1993) from the Colorado K/T section. There, different types of zircon from the boundary layer were dated by the U-Pb method, defining a lower intercept age of 65.5 ± 3 Ma in the concordia diagram. These data are reproduced in detail in Fig. 15. The zircon grains that plot close to the lower intercept have very peculiar crystal characteristics and it seems, that they re-crystallized in the hot vapor cloud of ejecta, producing strong lead loss in zircons derived from an about 550 Ma old target rocks. As illustrated by this case, the U-Pb dating method also allows to identify the target rocks, and in consequence, to search for the impact location, i.e. to correlate distant ejecta layers with their potential source craters.

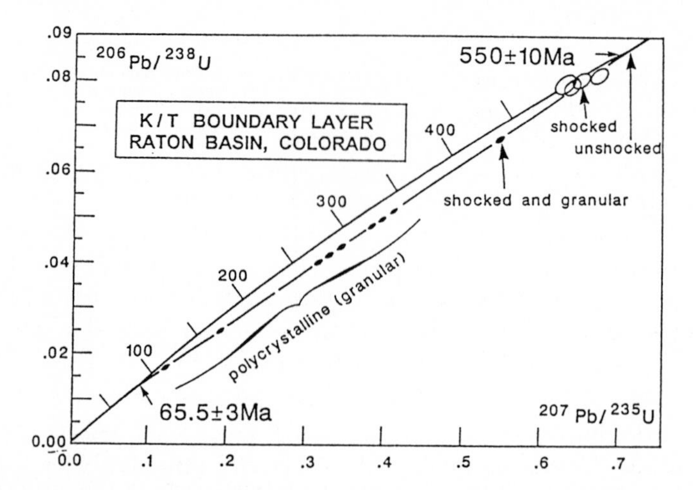

Fig. 15. Concordia diagram for zircons from the K/T boundary layer in Colorado, showing that originally 550 Ma old grains have experienced varying degrees of episodic Pb-loss during the impact event at 65.5 Ma (after : Krogh et al. 1993). Degrees of Pb-loss are correlated with increasingly shocked and re-crystallized grains.

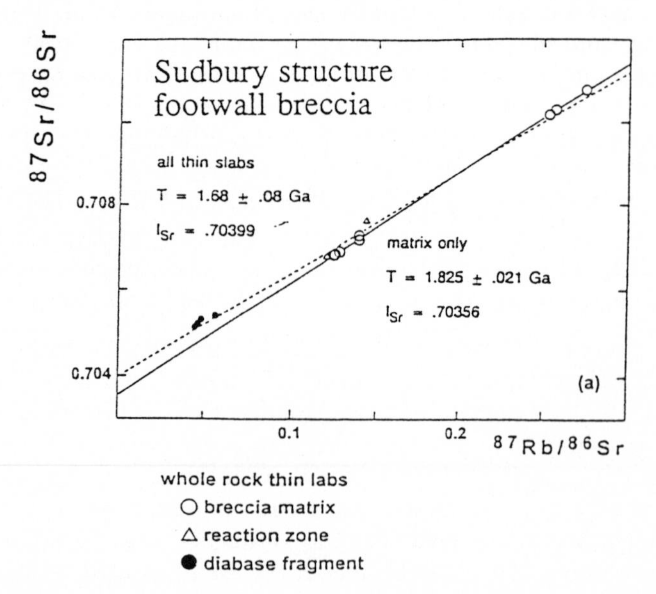

Fig. 16. Rb-Sr isochron for samples collected in the Sudbury Footwall Breccia that lies beneath the large impact melt sheet (after : Deutsch et al. 1989). Note that the diabase (basaltic dike) fragments are not in equilibrium with the matrix yielding a much younger age of 1.68 Ga. On the other hand, the 1.83 Ga age defined by the breccia matrix analyses is only slightly younger than the 1.85 Ga U-Pb age (Fig. 13). This difference could be due to the fact that long-lasting high temperatures lead to a slightly later closure (time extended diffusion) of the Rb-Sr system.

8.3 Crater Basement

A relatively reliable Rb-Sr age dating was achieved in the parauthochthonous Footwall Breccia lying directly beneath the Sudbury "Igneous" Complex. There, the large impact melt sheet that overlies the breccia caused annealing and melting of fine crushed matrix material, accompanied by a total re-homogenization of the Rb-Sr system in small domains of the matrix (Deutsch et al. 1989). Fig. 16 shows these data which yield an age of 1825 $\pm$ 21 Ma for different matrix samples. The isochron diagram also shows that small diabase fragments are not equilibrated with the embedding matrix giving a much younger age of about 1680 Ma. This age may be ascribed to later alteration processes affecting the breccia. As shown above, the precise impact age of 1850 Ma for the Sudbury event was determined by the U-Pb method (Krogh et al. 1984), and the slightly younger Rb-Sr matrix age of 1825 Ma can be explained by later closure of this system due to long-lasting solid state diffusion on the mm-scale. This observation suggests that cooling of deep crater levels may take several million years.

References

Allègre, C.J., Manhès, G., Göpel, C. (1995): The age of the Earth. Geochim. Cosmochim. Acta **59**, 1445–1456.

Baadsgaard, H., Lerbekmo, J.F., McDougall, I. (1988): A radiometric age for the Cretaceous-Tertiary boundary based upon K-Ar, Rb-Sr and U-Pb ages of bentonites from Alberta, Saskatchewan and Montana. Can. J. Earth Sci. **25**, 1088–1097.

Bogard, D.D. (1995): Impact ages of meteorites: A synthesis. Meteoritics **30**, 244–268.

Bottomley, R., York, D. (1988): Age measurement of the submarine Montagnais impact crater. Geophys. Res. Lett. **15**, 1409–1412.

Bottomley, R., York, D. (1991): The age of the North-American tectite event(s). In *GAC-MAC-SEG Meeting, Toronto, Program with Abstracts* **16**, A13.

Deutsch, A., Buhl, D., Langenhorst, F. (1992): On the significance of crater ages: new ages for Dellen (Sweden) and Araguainha (Brazil). Tectonophysics **216**, 205–218.

Deutsch, A., Lakomy, R., Buhl, D. (1989): Strontium and Neodynium isotopic characteristics of a heterolithic breccia in the basement of the Sudbury impact structure, Canada. Earth Planet. Sci. Lett. **93**, 359–370.

Deutsch, A., Schärer, U. (1990): Isotope systematics and shock-wave metamorphism: I. U-Pb in zircon, titanite, and monazite, shocked experimentally up to 59 GPa. Geochim. Cosmochim. Acta **54**, 3427–3434.

Deutsch, A., Schärer, U. (1994): Dating terrestrial impact events. Meteoritics **29**, 301–322.

Gibbins, W.A., McNutt, R.H. (1975): The age of the Sudbury nickel eruptive and the Murray granite. Can. J. Earth Sci. **12**, 1970–1989.

Grieve, R.A.F., Dence, M.R., Robertson, P.B. (1977): Cratering processes: As interpreted from the occurence of impact melts. In : Roddy, D.J., Pepin, R.O., Merill, R.B. (eds.), *Impact and Explosion Cratering*, Pergamon, p. 971–814.

Hall, C.M., York, D., Sigurdsson, H. (1991): Later 39Ar/40Ar step-heating ages from single Cretaceous-Tertiary boundary glass spherules. EOS **72**, No. 44 (supplement), p. 531.

Hartung, J.B., Izett, G.A., Naeser, C.W., Kunk, M.J., Sutter, J.F. (1986): The Manson, Iowa, impact structure and the Cretaceous-Tertiary boundary event (abstract). Lunar Planet. Sci. **18**, 313–314.

Izett, G.A., Dalrymple, G.B., Snee, L.W. (1991): 40Ar/39Ar age of the K/T boundary tectites from Haiti. Science **252**, 1539–1543.

Jahn, B.-M., Floran, R.J., Simonds, C.H. (1978): Rb-Sr isochron age of the Manicouagan melt sheet, Quebec, Canada. J. Geophys. Res. **83**, 2799–2803.

Jessberger, E.K. (1988): 40Ar/39Ar dating of the Haughton impact structure. Meteoritics **23**, 233–234.

Jessberger, E.K., Reimold, W.U. (1980): A late Cretaceous 40Ar/39Ar age for the Lappajärvi impact crater, Finland. J. Geophys. **48**, 57–59.

Krogh, T.E., Davis, D.W., Corfu, F. (1984): Precise U-Pb and baddeleyite ages for the Subdury area. In : Pye, E.G., Naldrett, A.J. Giblin, P.E. (eds.), *The Geology and Ore Deposits of the Sudbury Structure*, Ont. Geol. Surv. Spec. Paper 1, p. 431–446.

Krogh, T.E., Kamo, S.L., Bohor, B. (1993): Fingerprintring the K/T impact site and determining the time of impact by U-Pb dating of single shocked zircons from distal ejecta. Earth Planet. Sci. Lett. **119**, 425–430.

LaBreque, J.L., Kent, D.V., Cande, S.C. (1977): Revised magnetic polarity time scale for late Cretaceous and Cenozoic time. Geology **5**, 330–335.

Martinez, I, Schärer, U., Guyot, F. (1993): Impact-induced phase transformations at 50-60 GPa in continental crust: an EPMA and ATEM study. Earth Planet. Sci. Lett. **119**, 207–223.

Maurer, P., Eberhardt, P., Geiss, J., Grögler, N., Stettler, A., Brown, G.M., Peckett, A., Krähenbühl, U. (1978): Pre-Imbrian crater and basins: ages, compositions and excavation depths from Apollo 16 breccias. Geochim. Cosmochim. Acta **42**, 1687–1720.

McWilliams, M.O., Baksi, A.K., Bohor, B.F., Izett, G.A. and Murali, A.V. (1992): EOS **73**, No. 14 (supplement), p. 363.

Melosh, H.J. (1989): *Impact Cratering. A Geological Process*, Oxford Univ. Press.

Nishiizumi, K., Kohl, C.P., Shoemaker, E.M., Arnold, J.R., Klein, J., Fink, D., Middleton R. (1991): In situ 10Be-26Al exposure ages at Meteor Crater, Arizona. Geochim. Cosmochim. Acta **55**, 2699–2703.

Omar, G., Johnson, K.R., Hickey, L.J., Robertson, P.B., Dawson, M.R., Barnosky, C.W. (1987): Fission-track dating of Haughton astrobleme and included biota, Devon Island, Canada. Science **237**, 1603–1605.

Phillips, F.M., Zreda, M.G., Smith, S.S., Elmore, D., Kibik, P.W., Dorn, R.I., Roddy, D.J. (1991): Age and geomorphic history of Meteor Crater, Arizona, from cosmogenic 36Cl and 14C in rock varnish. Geochim. Cosmochim. Acta **55**, 2695–2698.

Schärer, U., Copeland, P., Harrison, T.M., Searle, M.P. (1990): Age, cooling history and origin of post-collisional leucogranites in the Karakoram Batholithe, N-Pakistan; a multi-system isotope study. *J. Geol.* **98**, 233–251.

Schärer, U., Deutsch, A. (1990): Isotope systematics and shock-wave metamorphism II: U-Pb and Rb-Sr in naturally shocked rocks; the Haughton impact structure. Geochim. Cosmochim. Acta **54**, 3435–3447.

Sharpton, V.L., Ward, P.D. (eds.) (1990): *Global Catastrophs in Earth History: An Interdisciplinary Conference on Impacts, Volcanism, and Mass Mortality*, Geol. Soc. Am. Spec. Paper 247, 631 pp.

Shoemaker, E.M. (1983): Asteroid and comet bombardment of the Earth. Rev. Earth Planet. Sci. **11**, 461–494.

Staudacher, T., Jessberger, E.K., Dominik, B., Kirsten, T., Schaeffer, O.A. (1982): 40Ar/39Ar ages of rocks and glasses from the Nördlingen Ries Crater and the temperature history of impact breccias. J. Geophys. Res. **51**, 1–11.

Stöffler, D. (1971): Progressive metamorphism and classification of shocked and brecciated crystalline rocks at impact craters. J. Geophys. Res. **76**, 5541–5551.

Suess, H.E., Hayden, R.L., Inghram, M.G. (1951): Age of tektites. Nature **168**, 432.

Sutten, S.R. (1985): Thermoluminescence measurements on shock-metamorphosed sandstone and dolomite from Meteor Crater, Arizona. J. Geophys. Res. **90**, 3690–3700.

Swisher III, C.C., Grajales-Nishimura, J.M., Montanari, A., Margolis, S.V., Claeys, P., Alvarez, W., Renne, P., Cedillo-Pardo, E., Maurasse, F.J.-M. R., Curtis, G.H., Smit, J., Mc Williams, M.O. (1992): Coeval 40Ar/39Ar ages of 65.0 million years ago from Chixculub crater melt rocks and Cretaceous-Tertiary boundary tektites. Science **257**, 954–958.

Taylor, S.R. (1982): *Planetary Sciences : a Lunar Perspective*, Lunar and Planetary Institute, Houston, U.S.A.

Torigoye, N., Shima, M. (1993): Evidence for a late thermal event of unequilibrated enstatite chondrites: A Rb-Sr study of Qingzhen and Yamato 6901 (EH3) and Khairpur (EL6). Meteoritics **28**, 151–527.

Vogt, S., Herzog, G.F., Reedy, R.C. (1990): Cosmogenic nuclides in extraterrestrial materials. Rev. Geophys. **28**, 3, 253–275.

Wagner, G.A., Van den haute, P. (1992): *Fission-track dating*, Ferdinand Enke Verlag, Stuttgart, Germany.

Impact Energy Flux on Earth in the Last 150 Ma as Inferred from the Cratering Records

Adriano Campo Bagatin[1,4], Alessandro Montanari[2,3] and Paolo Farinella[1]

[1] Dipartimento di Matematica, Università di Pisa
Via Buonarroti 2, I - 56127 Pisa, Italia
[2] Osservatorio Geologico di Coldigioco, I - 62020 Frontale di Apiro, Italia
[3] Ecole des Mines de Paris, France
[4] Departamento de Matematica Aplicada i Astronomia, Universitat de Valencia
Dr. Moliner 50, E - 46100 Burjassot (Valencia), España

Flux énergétique d'impact sur Terre durant les derniers 150 millions d'années, établi d'après les comptages de cratères

Résumé. Notre échantillon est constitué de 30 grands (c'est-à-dire de diamètre supérieur à 5 km) cratères d'impact terrestres dont les âges sont bien connus; tous sont plus jeunes que 150 millions d'années. A partir de leurs dimensions et de leurs âges, ainsi que des incertitudes sur ces valeurs, nous avons construit un graphe résumant notre connaissance actuelle sur le flux énergétique d'impact sur la Terre en fonction du temps.
Le diamètre.d'un cratère indique l'énergie de l'impact qui lui a donné naissance, grâce à des lois d'échelle convenables; on lui associe alors une fonction du temps gaussienne. Toutes les fonctions des différents cratères sont sommées, et la courbe résultante a été lissée par moyennisation par intervalles de 4 millions d'années.
Il en ressort que le cratère de Chicxulub, vieux de 65 millions d'années et associé à la transition crétacé-tertiaire, correspond au plus fort pic de flux d'énergie; ce pic dépasse de presque un ordre de grandeur le deuxième. D'autre part, on en déduit que la dimension minimale d'un impacteur, pour causer des extinctions massives d'espèces, se situe vers 3 km environ.
Bien qu'il n'y ait pas d'évidence convaincante de périodicité dans la distribution des âges des cratères, quelques groupes de cratères semblent être plus proches en âge que ce que donnerait une distribution au hasard.

Abstract. We have used a compilation of 30 well-dated large impact craters on Earth (i.e., diameters larger than 5 km) younger than 150 Ma, their diameters, geochronologic ages, and the corresponding uncertainties to construct a graph summarizing our current knowledge on the influx of the impact energy onto the Earth as a function of time. From the crater diameters, we estimated the corresponding impact energies through suitable scaling laws. Then to each crater we associated a gaussian (bell) function of time centered at its age. Finally, all the bell functions corresponding to different craters were summed up and the resulting curve was plotted. From this curve, it is apparent that the 65 Ma old Chicxulub crater associated with the Cretaceous/Tertiary boundary corresponds to the highest energy influx peak, almost an order of magnitude larger than the next ones, and that there is probably a threshold size of $\approx$ 3 km for the smallest projectile capable of triggering large-scale extinctions. Although there is no convincing evidence for periodicities in the distribution of the crater ages, a few group of several craters appear to be more closely spaced in time than in a purely random distribution.

1 Introduction

The main purpose of this chapter is to review the Earth's cratering record and the biological effects of the impacts undergone by our planet in the (geologically) recent past. In particular, we will try to estimate the energy flux associated with the greatest cratering events that happened on the Earth in the last 150 Ma (Ma = mega annum = 10^6 years), as a function of time, to infer from the available cratering record any possible evidence for periodicities or other non–random features in the distribution of such events, and to analyse their correlation with global environmental catastrophes inferred from the geologic and paleontologic record.

We have thus compiled a list of the largest impact cratering events occurred in the last 150 Ma by considering all the craters observed on Earth whose diameter is greater than 5 km. Actually, there are more than 45 known craters with this size, but we have taken into account only 30 of them because, as we are interested in their time distribution, we needed fairly precise and accurate estimates of crater ages with a 2σ error $<$ 10 Ma. This selection results in the record of craters listed in Table 1 (note that they are numbered up to 29 because event no. 19 corresponds to a doublet crater, with a single age value).

Figure 1 shows the distribution of the selected craters plotted against a chronostratigraphic time scale, and correlated with the curves of extinction of marine organism families (Raup and Sepkosky 1986), and genera (Sepkosky 1990), and the signatures of impacts including iridium anomalies, shocked quartz, spherules (i.e., microtektites and microcrystites), megawave deposits, and anoxic crises recorded in sedimetary sequences. In the next section, we will examine the reliability of the complex set of data upon which our subsequent impact energy flux analysis is based.

2 The Cratering Record and Its Uncertainties

The most crucial aspect of Fig. 1 is the geochronologic time frame on which the correlation between stratigraphic signatures, paleontologic record, and impact craters is based. This is to say that the ejecta of an impact found in a sedimentary layer, and a biologic signature which may be caused by an impact, have to be dated in terms of numerical age in Ma in order to be correlated with a known impact crater. The only way to estimate the age of a crater is to date its impact melt rock with a radioisotopic method. In fact, most of the craters are found on land and are buried or filled by incomplete sequences of terrestrial sediments which do not permit a direct stratigraphic correlation with marine, fossiliferous sediments. On the other hand, marine sedimentary sequences are dated by interpolation of few radioisotopic ages, obtained from interbedded volcanic ashes in distant sections around the world, and tied to paleontological, geochemical, and geophysical signatures which are assumed to be contemporaneous worldwide. Therefore, the strength of a correlation between an impact signature found in the sedimentary record and an apparently contemporaneous crater is solely based on the accuracy and precision of the geochronologic calibration of the time scale, and the reliability of the geochronologic date of that crater.

For the calibration of the Tertiary time scale (from the top of the Cretaceous to the Pliocene, see Fig. 1), we used the remarkable record of the Umbria-Marche pelagic sequence, which is extensively exposed throughout the northeastern Appennines of Italy. This is a unique situation, where a continuous and complete sequence of fossiliferous marine sediments contain numerous volcanic ashes which permit direct radioisotopic calibration of the chrono–stratigraphic time scale (e.g., Montanari et al. 1988, 1991). Moreover, the Umbria-Marche sequence contains the signatures of impacts (i.e., spherules, shoched quartz, Ir anomalies, biologic crises) at the Cretaceous (K/T) boundary (e.g. Alvarez et al. 1980; Montanari et al. 1983; Montanari 1991) and in the Late Eocene (e.g., Montanari et al. 1993; Clymer et al. 1995), which appear to be correlatable in numerous other sections worldwide. As for the geochronologic calibration of the Cretaceous Period, we have used the time scale of Harland et al. (1989).

A remarkable aspect of Fig. 1 is the obvious scarcity of impact signatures known at present in the stratigraphic record, compared to the number of large craters on the surface of the Earth. This is mainly due to the fact that the search for signatures such as Ir anomalies, spherules, and shocked minerals, has been mainly focused across those few short stratigraphic intervals where major extinctions are recorded, or which were known to cover the time of major impacts. The job of the stratigrapher searching for a millimetric impact layer in a sedimentary sequence hundreds of meters thick is comparable to the proverbial search for the needle in the haystack.

Fig. 1. *(facing page).* Impact craters and extinctions plotted against the stratigraphic column of the Umbria–Marche sequence. The main extinctions are: MM/LM = Middle/Late Miocene boundary; LE = Late Eocene (E/O) boundary; K/T = Cretaceous/Tertiary (KT) boundary; C/T = Cenomanian/Turonian boundary; A/A = Albian/Aptian boundary; H/V =3D Hauterivian/Valanginian boundary; B/T = Berriasian/Tithonian boundary. Extinction intensities have been estimated after Raup and Sepkoski (1986) and Sepkoski (1990). The size and distribution of impact craters through time is mainly derived from Grieve (1991) and Hodge (1994).

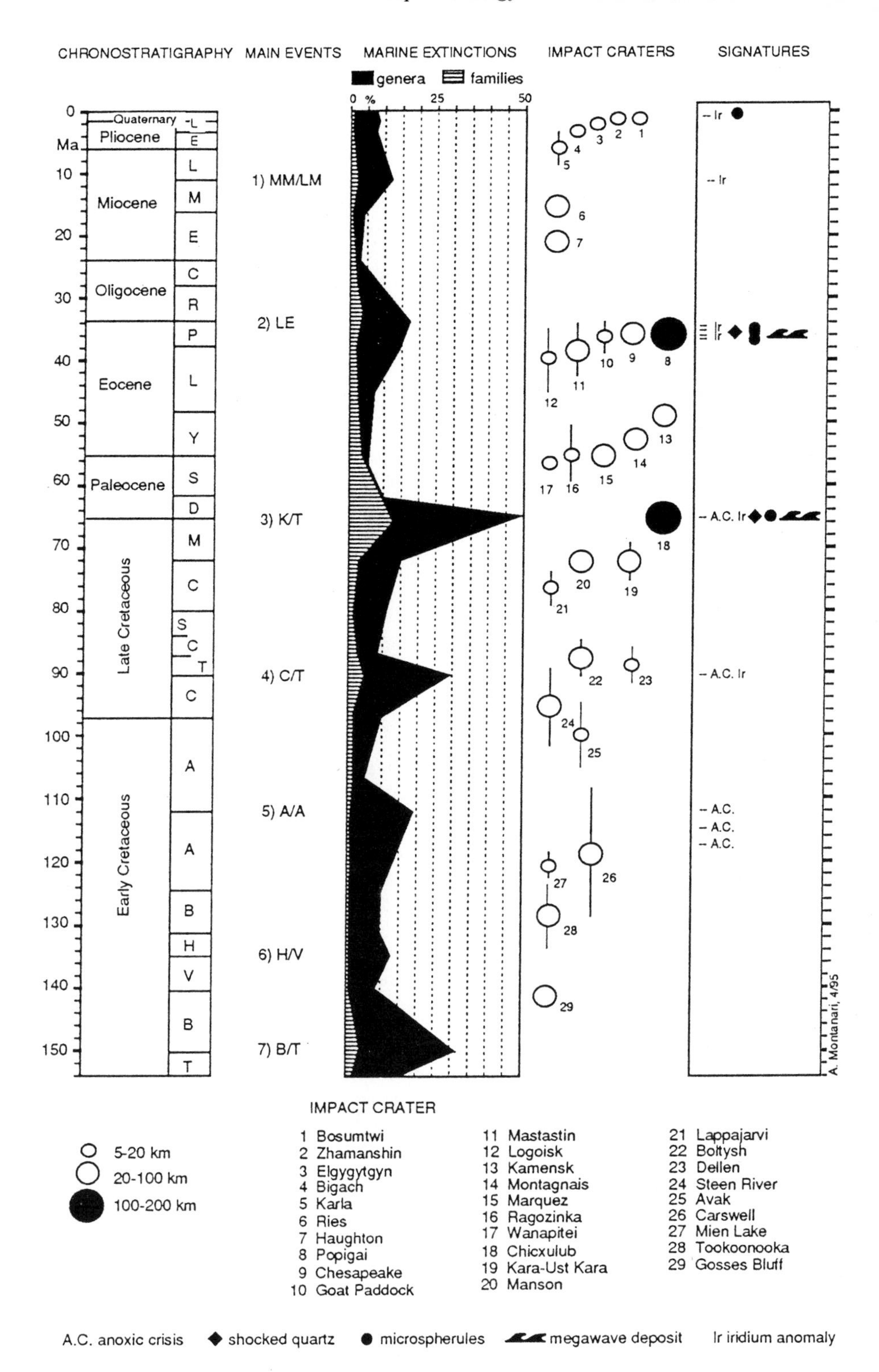
CHRONOSTRATIGRAPHY
MAIN EVENTS
MARINE EXTINCTIONS
IMPACT CRATERS
SIGNATURES
genera
families
0 %
25
50
Ma
Quaternary
Pliocene
Miocene
Oligocene
Eocene
Paleocene
Late Cretaceous
Early Cretaceous
1) MM/LM
2) LE
3) K/T
4) C/T
5) A/A
6) H/V
7) B/T
Ir
A.C. Ir
A.C.
A. Montanari, 4/95
IMPACT CRATER
5-20 km
20-100 km
100-200 km
1 Bosumtwi
2 Zhamanshin
3 Elgygytgyn
4 Bigach
5 Karla
6 Ries
7 Haughton
8 Popigai
9 Chesapeake
10 Goat Paddock
11 Mastastin
12 Logoisk
13 Kamensk
14 Montagnais
15 Marquez
16 Ragozinka
17 Wanapitei
18 Chicxulub
19 Kara-Ust Kara
20 Manson
21 Lappajarvi
22 Boltysh
23 Dellen
24 Steen River
25 Avak
26 Carswell
27 Mien Lake
28 Tookoonooka
29 Gosses Bluff
A.C. anoxic crisis
shocked quartz
microspherules
megawave deposit
Ir iridium anomaly

Table 1. List of impact craters on Earth younger than 150 Ma and larger than 5 km in diameter.

Crater name and locality	diameter (km)	age (Ma)	error (2σ)
1) Bosumtwi (Ghana)	10.5	1.30	0.10
2) Zhamanshin (Kazakhstan)	13.5	0.90	0.02
3) Elgygytgyn (Russia)	18.0	3.50	0.50
4) Bigach (Kazakhstan)	7.0	6.00	3.00
5) Karla (Russia)	12.0	10.00	5.00
6) Ries (Germany)	24.0	14.80	0.70
7) Haughton (Canada)	24.0	23.40	1.00
8) Popigai (Russia)	100.0	35.60	0.80
9) Chesapeake (U.S.A.)	85.0	36.00	2.00
10) Goat Paddock (Australia)	5.1	37.00	3.00
11) Mistastin (Canada)	28.0	38.00	4.00
12) Logoisk (Russia)	17.0	40.00	5.00
13) Kamensk (Russia)	25.0	49.20	0.20
14) Montagnais (Canada)	45.0	50.50	0.76
15) Marquez (U.S.A.)	15.0	58.30	3.10
16) Ragozinka (Russia)	9.0	55.00	5.00
17) Wanapitei (Canada)	7.5	57.00	2.00
18) Chicxulub (Mexico)	200.0	64.98	0.05
19a) Kara (Russia)	65.0	73.00	3.00
19b) Ust Kara (Russia)	25.0	73.00	3.00
20) Manson (U.S.A.)	35.0	73.80	0.30
21) Lappajärvi (Finland)	23.0	77.00	0.40
22) Boltysh (Ukrainia)	24.0	88.00	3.00
23) Dellen (Sweden)	15.0	89.60	2.70
24) Steen River (U.S.A.)	25.0	95.00	7.00
25) Avak (U.S.A.)	12.0	100.00	5.00
26) Carswell (Canada)	39.0	115.00	10.00
27) Mien Lake (Sweden)	9.0	121.00	2.30
28) Tookoonooka (Australia)	55.0	128.00	5.00
29) Gosses Bluff (Australia)	22.0	142.00	0.50

It has to be stressed that the record shown in Table 1 and Fig. 1 does not represent all the impact events occurred on Earth in the past 150 Ma, but only the few ones that have left their signature on the Earth's surface and have been discovered up to now. There are at least three kinds of problems which limit the accuracy for a detailed statistical study of the impact record through Earth history. Here is a brief review of them.

A first source for analytical inaccuracy lies in the actual incompleteness of the cratering record due to the fact that the Earth's surface is extremely dynamic on a geologic time scale: many craters may have been completely obliterated by erosion or tectonic deformation, or buried under orogens, or subducted under tectonic plates. Moreover, the size vs. frequency relationship of impacts follows approximately a power law: small events (i.e. capable of excavating a crater 20 km in diameter or less) are much more frequent than medium size events (20 to 100 km in diameter). In turn, medium size events are much more frequent than giant impacts producing craters larger than 100 km. On the other hand, the preservation probability of impact craters grows with their size: small structures may be erased from the record much more easily than large ones. Consequently, the record in Fig. 1 and Table 1 shows an abnormous abundance of medium size craters (averaging 25 km in diameter), and an anomalous scarcity of small size craters which, according to a power law distribution, should be much more numerous.

This observation also suggests that the probability that a giant (> 100 km) crater to be obliterated and thus unrecorded in the examined time interval is lower, but still significant. Nevertheless, medium or large impact events are the ones that count more in the general study of impact flux and possible cause-and-effect relationship between impacts and biologic crises. Large impacts are those which may leave a world-wide signature in the geologic record, and may have climatic, environmental, and biologic effects on a global scale.

A second problem lies in the fact that many impact craters have been discovered recently, at a rate of approximately 1-2 new ones per year. This is due to the fact that on land there are still vast areas such as the poles, deserts and tropical forests, which have not yet been thoroughly explored for impact structures. Moreover, about 2/3 of the Earth's surface is made of oceanic crust, and at present all the known impact craters are located on continental crust.

Just to make an example of how critical the incompleteness of the record may be, let us consider the pioneering work by Alvarez and Muller (1984), which analyzed the cratering periodicity in the past 250 Ma. This study was based on a record of only 13 craters larger than 10 km in diameter. In the present study, we have analyzed the distribution of 25 craters with diameters larger than 10 km, and through a much shorther geologic time span (150 Ma) than that analyzed by Alvarez and Muller. The recent flux of new entries in cratering records may explain, by statistical arguments, the discrepancies between different investigations (for instance on periodicities) that have been carried out in the course of the past decade (e.g., Alvarez and Muller 1984; Pohl 1987; Baksi 1990). A study with a few tens of craters which seem to fit well a periodic age distribution, may be invalidated merely by introducing a number of new craters into the record.

Finally, there are still basic geochronologic uncertainties in the age estimate of impact craters.

This study, as well as preceding studies of this kind, is strongly dependant on the reliability of radioisotopic dates of craters, and the chronostratigraphic time scale which, as geochronology goes, improves with the progress of sophistication of analytical instruments and methods. For instance, for a long time the medium size Manson crater was believed to have an age close to the K/T boundary (i.e., about 66 Ma; Hartung et al. 1990). Recently, the melt rock of this crater has been re-dated yielding an age about 7 Ma older (Izett et al. 1995). Similarly Kamensk, another medium size crater (see Table 1) was for long time considered a K/T boundary event and only recently was it re-dated at 49.2 Ma (i.e., 16 Ma younger; Izett et al. 1994). Another relevant case is the Kara-Ust/Kara double structure, which was dated by a Russian team using the traditional K/Ar technique at around 66 Ma (Kolesnikov et al. 1988), and in an American geochronology laboratory using the Ar^{40}/Ar^{39} technique at 75 Ma (Koeberl et al. 1990). The latter date is herein the preferred one. As for the crater size, the diameter of the Kara crater, which is entirely located on land, is fairly well established at 65 km, whereas the size of the Ust Kara, submerged under the Kara Sea, is still uncertain. Early estimates indicated a diameter for the Ust Kara of about 25 km; further geological studies by Nazarov et al. (1989) suggested a diameter between 70 km and 155 km, whereas analyses of geophysical prophiles led Koeberl et al. (1990) to a diameter estimate of 80 km. Here we have adopted the smallest estimate, but this may well turn out to be wrong in the future.

In summary, we have conducted this study mainly to analyze as simply and objectively as possible the time distribution of the 30 best known large craters on Earth throughout the best known interval of Earth history (i.e., the past 150 Ma), out of an inferred total population of perhaps a thousand impact craters which are unknown to us because not yet discovered or which have been completely obliterated by geological processes. From this partial record, we are going to deduce the energy distribution vs. time of the impacts that caused the known craters, and to search for a possible periodicity of these events.

3 Impact Energy Flux Analysis

3.1 Cratering Energy Scaling

Given the craters listed in Table 1, we have now to estimate the energy associated to any single cratering impact, namely the kinetic energy of the corresponding extraterrestrial projectiles (either comets or asteroids). Since large–scale impacts have never been directly observed on Earth, and the evidence provided by experiments (including nuclear tests) is limited to energies lower than those of km–sized extraterrestrial impactors by many orders of magnitude, we have to resort to suitable *scaling laws*. These are extrapolating relationships inferred from experiments through dimensional analysis, which allow one to relate the outcome of an impact (e.g., the characteristics of the crater) to the physical properties of the projectile.

A number of such relationships have been proposed over the years for cratering impacts (see Melosh 1989, Sect. 7.8). We have adopted Gault's (1974) scaling law, to be applied to lunar craters larger than ≈ 1 km across:

$$D_{at} = 0.27\, \rho_p^{1/6} \rho_t^{-1/2} W^{0.28} (\sin\theta)^{1/3}\,, \qquad (1)$$

where D_{at} stands for the "apparent" crater diameter, as shown in Fig. 2, ρ_p and ρ_t are the projectile's and target's densities, respectively, W is the projectile's kinetic energy (all these quantities being expressed in SI units), and θ is the impact angle with respect to the local zenith. We have chosen this scaling law among the several ones discussed by Melosh essentially for the sake of simplicity, as it does not involve additional unknown parameters such as the projectile's size. As emphasized by Melosh, owing to the differences between the predictions of different scaling laws, the determination of W from the crater diameter is rather imprecise, with about one order of magnitude of uncertainty for large craters. However, since we are interested only in determining relative (rather than absolute) impact energies over a limited range of crater sizes, and since there are other significant sources of uncertainty (to be discussed below), our results are not really sensitive to the chosen scaling relationship.

In order to apply Eq. (1) to the observed diameters of terrestrial craters, two corrections are needed. First, we have to account for the weaker lunar gravity. First, according to Gault the crater's diameter scales with the -0.165 power of the surface gravity acceleration, so the same projectile will generate on the Earth a crater $(9.8/1.67)^{-0.165} = 0.75$ times as big as on the Moon. Second, the "apparent" diameter does not correspond to the visible (i.e. measurable) *rim to rim* diameter (see Fig. 2). For this, we shall adopt Holsapple's (1993) estimate that the measurable crater diameter D_0 is $\approx 1.3 \times D_{at}$. Applying these two corrections and inverting Eq. (1) to obtain W, we have

$$W = 123\, \rho_p^{-0.595} \rho_t^{1.786} (\sin\theta)^{-1.19} D_0^{3.571} \qquad (2)$$

(SI units). This is the scaling relationship we will use in the following to derive impact energies from crater diameters.

As for the other parameters appearing in Eq. (2), we adopted the following numerical values: $\rho_p = 2500$ kg/m^3, in agreement with the estimated densities of asteroids, $\rho_t = 3500$ kg/m^3 for the Earth's crust, and $\theta = 60°$ (the median value for an isotropic flux of impactors). Owing to the different exponents of the power laws making up Eq. (2), it is easy to show that a 10% uncertainty in the crater diameters corresponds to a larger error in the estimated energy than due to the expected uncertainty in the densities ($\approx 20\%$, assuming that only a minor fraction of craters are due to comet impacts). Taking into account the unknown impact angles, it is clear that for individual cratering events our W estimates may well be wrong by $\pm$ a factor of 5. However, this unavoidable uncertainty is not likely to affect the main qualitative conclusions we are going to draw from the present analysis.

Of course all the steps of this procedure are somewhat arbitrary, and may be replaced by alternative ones, yielding quantitatively different results. However, given all the sources of uncertainty we discussed earlier, we are mainly interested in the qualitative features of the curve shown in Fig. 3. These features, to be discussed in Sect. 4, are not really sensitive to the details of the derivation procedure.

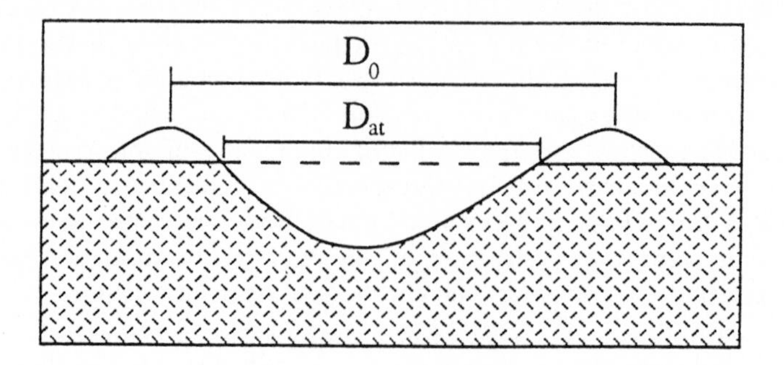

Fig. 2. Sketch showing the difference between the apparent diameter (D_{at}) and the true rim–to–rim diameter (D_0) of an impact crater.

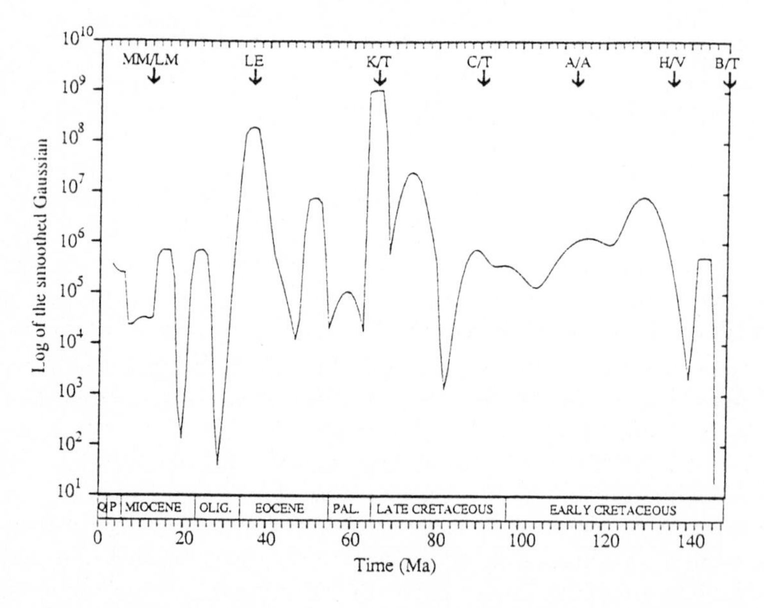

Fig. 3. A semi–logarithmic diagram of impact energy flux vs. time over the last 150 Ma (see text). Major biological crises (as in Fig. 1) are indicated by arrows.

3.2 Energy Flux

In order to have an idea of how the impacting bodies deposited their energy as a function of time, we have chosen the following methodology. For every impact we considered the diameter of the observed crater, its age estimate and the corresponding 2σ error. Then we associated to every crater a gaussian (bell–like) distribution centered on its estimated age, with a half–width equal to σ (half of the assumed error on the age estimate), and such that its total area is normalized to the estimated energy W of the corresponding cratering event – according to the scaling law described in Sect. 3.1. Then, we summed up all the gaussian curves and represented the resulting diagram on a semilogarithmic scale for energy flux vs. time (the energy deposition rate is measured in Megaton/Ma, where 1 Megaton is the energy released by 10^9 kg of TNT explosive, i.e., about 4.2×10^{15} J). The logarithmic energy scale was more suitable than a linear one because of the large differences (up to 5 orders of magnitude) between the energies associated with craters of different sizes. To smooth out the sharpest peaks of the curve generated in this way, we computed running averages over a succession of 4 Ma interval, and plotted in Fig. 3 the resulting smoothed curve.

3.3. Distribution of Impact Times

The time distribution of the cratering events included in our sample can be also analysed, in order to look for possible non–random features. To check for possible periodicities, we used a simple test based on the distribution of crater age differences, devised by S. Perlmutter and applied by Alvarez and Muller (1984) with a more limited set of data. This test consists in computing all the time differences between any pair of dated events, and then representing the frequency with which any given value occurs in the data, as a function of the time difference itself. Every difference is plotted as a gaussian with the two errors of the corresponding ages combined quadratically, and all the gaussians are superimposed. In this way we get a curve with a number of peaks corresponding to the most frequent time differences. It is clear that a periodic signal in the data would yield peaks at time differences equal to the period and its multiples. A more refined Fourier analysis is warranted if from this test there is some hint for a periodicity.

The results of the test in the case of the crater age data listed in Table 1 is shown in Fig. 4. Although some peaks are apparent in this curve, they do not occur for multiples of a single value, and therefore are not likely to point to any significant periodic signal in the data. In order to show that the origin of these peaks is rather related to random fluctuations due to small number statistics, we have applied the same procedure with 10 sequences of randomly generated sets of time differences with the same total number of events. In Fig. 5, to the curve of Fig. 4 (the dashed one here) we have superimposed the 10 curves obtained with the random sets. The curve corresponding to the real data does not stand out in any significant way when compared to the random simulations. Actually, in some cases the random data yield peaks which are much higher than those of Fig. 4, and despite this no periodicity is obviously involved. Therefore, this simple test is enough to conclude that the crater age data do not provide any significant empirical evidence for a periodic component of the Earth impact flux, confirming a number of previous analyses of the same problem (for a recent review, see e.g. Grieve and Shoemaker 1994).

The fact that no periodicity is present in the data does not necessarily mean that they are completely random, and that all the impacts are to be seen as uncorrelated events distributed according to Poisson statistics. Actually, from Fig. 1 some "clusters" of craters closely spaced in time are quite apparent, and one may wonder whether they are statistically significant or just random flukes. To answer this question, one can apply a simple Poisson test. According to the

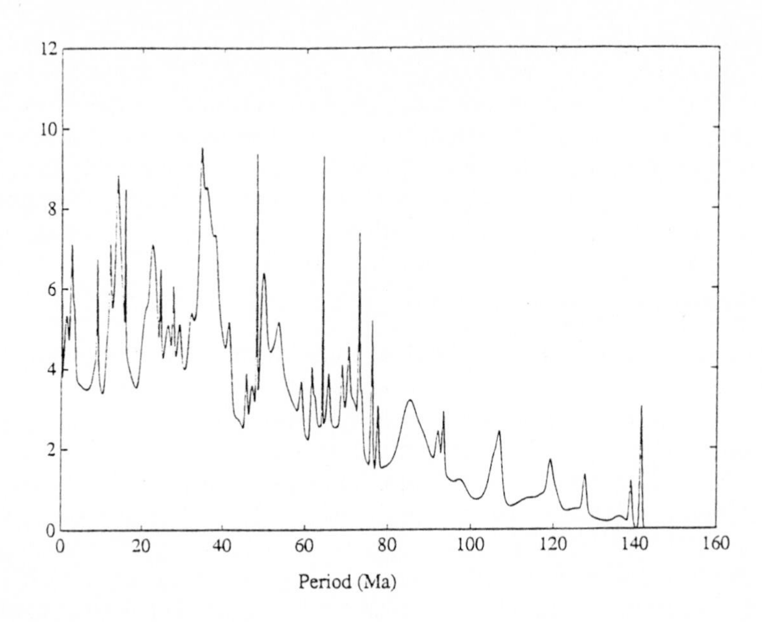

Fig. 4. Frequency of time difference values for all the pairs of cratering events from Table 1 (Perlmutter periodicity test, see text).

Poisson distribution, for a random sequence of events the probability of finding N events in a time interval where $\overline{N}$ are expected on average is

$$P(N) = (\overline{N}^N e^{-\overline{N}})/N! \qquad (3)$$

Applying this formula to the "clusters" of craters 1 to 5, 8 to 12, 13 to 17 and 19 to 21 (numbers as in Fig. 1 and Table 1), assuming average rates of $13/50 = 0.26$, $12/50 = 0.24$ and $4/50 = 0.08$ events per Ma in the three intervals between 0 and 50, 50 and 100, 100 and 150 Ma ago, respectively (see Table 1), we obtain that all the four "clusters" had an *a priori* probability of occurrence $< 10\%$ if the process was a random one. The Late Eocene craters 8 to 12, concentrated between 35 and 40 Ma ago, yield an *a priori* probability of only 0.8%. Thus we have some evidence for correlated groups of impacts occurring over intervals of several Ma. In other words, the Earth has undergone a significantly variable cratering rate, with enhancements of the order of a factor 4 for projectile sizes of 1 km and larger. This finding is not surprising, given our current understanding of the physical and dynamical evolution of comets and asteroids: an enhanced Earth cratering rate over several Ma could result either from discrete collisions in the main asteroid belt, injecting swarms of km–sized fragments into "fast–track" chaotic *routes* to the Earth–crossing region (Farinella et al. 1993, 1994; Menichella et al. 1996), or from the disintegration of sizeable comet nuclei, possibly after a Sun–grazing passage (Bailey et al. 1994).

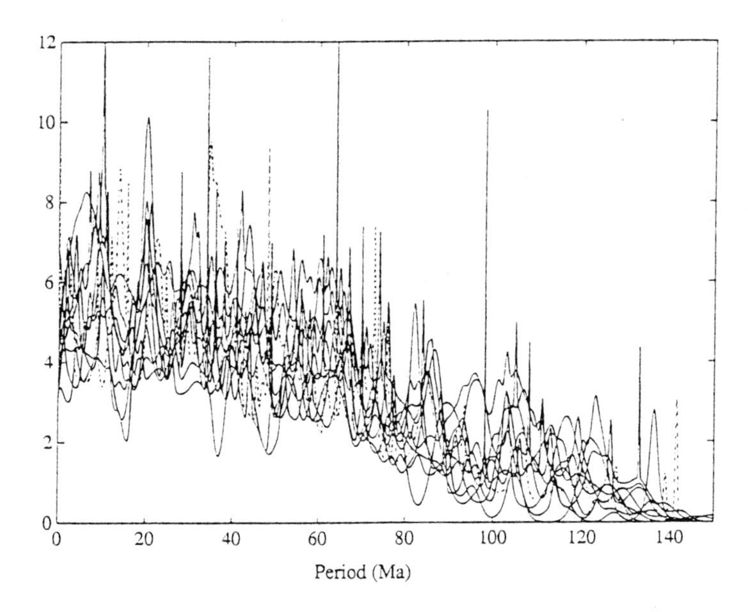

Fig. 5. The same as Fig. 4, but superimposing 10 curves obtained from fictitious random sequences to the (dashed) curve corresponding to the real data.

4 Discussion

The results reported in Sect. 3 have to be weighed with the comments made in Sect. 2. The incompleteness of the impact record as a whole and the scarcity of impact signatures in sedimentary sequences limit, perhaps inesorably, an accurate estimate of the overall flux of energy delivered to the Earth by colliding extraterrestrial objects. Nevertheless, a few deductions can be drawn from this study. For instance, no evidence for periodicity emerges from the processing of the data considered in this work. As stated above, the set of data we have analized may represent just a few percent of all the impact craters > 5 km in diameter that were formed on Earth in the past 150 Ma. There is no way to infer the time distribution of hundreds of other missing events. However, it can be said that the record of large impacts capable of escavating giant craters, which are more resistant to geologic obliteration, is closer to the unknown total number of this category of events than the few small craters (< 20 km in diameter) known in the studied interval of geologic time. In any case, because of the way impact energy scales with crater diameter, the energy flux vs. time diagram shown in Fig. 3 gives a disproportionate weight to large craters, and no obvious periodic feature is seen in this diagram too.

As far as the correlation of paleontological extinctions with the energy distribution associated with extra–terrestrial impacts is concerned, encouraging but not decisive arguments can be deduced by comparing Fig. 1 with Fig. 3. As shown in Fig. 3, the energy flux associated with the Chicxulub giant crater (some 200 km in diameter), which has been accurately and precisely dated at 64.98 ± 0.05

Ma (Swisher et al. 1993) and probably was the cause of the mass extinction at the K/T boundary (the most prominent in the analyzed time interval), corresponds by all means to the highest peak. The second energy peak shown in Fig. 3, almost one order of magnitude lower, is the result of five closely spaced impacts which occurred in the Late Eocene, and correlates well with another prominent extinction peak. Among the Late Eocene impacts, the Chesapeake (85 km in diameter) and the Popigai (100 km in diameter) are, after Chicxulub, the largest craters in the analyzed record. As for the rest of the 150 Ma record, there seems to be weak or no relation between other extinction maxima and impact energy peaks, which are anyway orders of magnitude smaller than the K/T and Late Eocene events. The sequence of closely spaced small and medium size impacts in the Early Eocene actually corresponds to a minimum in extinction rate (see Fig. 1). The Kara–Ust/Kara doublet crater, which may represent a very energetic and still uncertain event in the Late Cretaceous (Koeberl et al. 1990), did not leave a significant signature in the paleontologic record. Moreover, it is worth pointing out that Ir anomalies, or other impact signatures in the stratigraphic record, do not necessarily indicate large impacts capable of affecting the global biota. The small Ir anomaly found in marine sediments near the Middle/Late Miocene boundary (Asaro et al. 1988) is approximately correlated with a prominent extinction peak, but there is no known large crater associated with these signatures. On the other hand, the group of closely spaced small impacts in the Pliocene are not associated with any relevant biologic crisis.

These observations lead to the view that one single giant impact, such as the Chicxulub event, may cause a significant global biologic crisis, whereas smaller impacts, even if distributed in a geologically short span of time, may have no relevant effects on the global environment and biota recordable in the stratigraphic sequence. In other words, there may be an actual lower impact threshold (in terms of impactor size and energy release), above which a global environmental catastrope may occur. Such a threshold effect was proposed by Jansa et al. (1990), who deduced as zero extinction threshold limit the impact caused by a 3 km object (i.e., the Montagnais crater with a diameter of 45 km), and about 50% genera extinction with a 10 km bolide impact (i.e., the Chicxulub crater with a diameter of 200 km). Our results and observations reported in this paper agree with this empirical inference.

As for the time distribution of impact events, our analysis supports previous results on the absence of any convincing evidence for periodicities. On the other hand, the Late Eocene and possibly a few other groups of several craters appear more closely spaced in time than in a purely random distribution, pointing to the intriguing possibility that discrete break–up events of sizeable parent comets or asteroids generated swarms of Earth–crossing fragments, yielding significant enhancements of the Earth impact rate over intervals of 5–10 Ma.

In conclusion, there is still considerable work to be carried out for an acceptable understanding of the cause–effect relationship between large impacts and global biologic crises. This can be developed both by searching new craters hidden somewhere throughout the Earth surface, and with further systematic high–

resolution interdisciplinary studies of the stratigraphic impact record in continuous and complete marine sedimentary sequences. In doing this, the causal relationship between impacts and biological crises can be taken as a simple working hypothesis, and high–resolution stratigraphic analyses should be focused across those geologic time intervals where large impacts are known to have occurred, and/or where major extinction events are in some way recorded.

References

Alvarez, L.W., Alvarez, W., Asaro, F., Michel, H.V. (1980): Extraterrestrial cause for the Cretaceous-Tertiary extinction. Science **208**, 1095–1108.

Alvarez, W., Muller, R.A. (1984): Evidence from crater ages for periodic impacts on the Earth. Nature **308**, 718–720.

Asaro, F., Alvarez, W., Michel, H.V., Alvarez, L.W., Anders, M.H., Montanari, A., Kennett, J.P. (1988): Possible world-wide Middle Miocene iridium anomaly and its relationship to periodicity of impacts and extinctions. Lunar Plan. Institute **673**, 6–7.

Bailey, M.E., Clube, S.V.M., Hahn, G., Napier, W.M., Valsecchi, G.B. (1994): Hazards due to giant comets: Climate and short-term catastrophism. In Gehrels T. (ed.) *Hazards due to comets and asteroids*, Univ. of Arizona Press, pp. 479–533.

Baksi, A.K. (1990): Search for periodicity in global events in the geologic records: Quo vadimus? Geology **18**, 983–986.

Clymer, A.K., Bice, D.M., Montanari, A. (1995): Shocked quartz from the late Eocene: Impact evidence from Massignano, Italy. Geology **24**, 483–486.

Farinella, P., Gonczi, R., Froeschlé, Ch., Froeschlé, C. (1993): The injection of asteroid fragments into resonances. Icarus **101**, 174–187.

Farinella, P., Froeschlé, Ch., Froeschlé, C., Gonczi, R., Hahn, G., Morbidelli, A., Valsecchi, G.B. (1994): Asteroids falling into the Sun. Nature **371**, 314–317.

Gault D.E. (1974): Impact cratering. In Greeley R., Schultz P.H. (eds.) *A primer in lunar geology*, NASA Ames, Moffett Field, pp. 137–175.

Grieve, R.A.F. (1991): Terrestrial impact: the record in the rocks. Meteoritics **26**, 174–194.

Grieve, R.A.F., Shoemaker, E.M. (1994): The record of past impacts on Earth. In Gehrels T. (ed.) *Hazards due to comets and asteroids*, Univ. of Arizona Press, pp. 417–462.

Harland, B.W., Armstrong, R.L., Cox, A.V., Craig, L.E., Smith, A.G., Smith, D.G. (1990): *A geologic time scale 1989*, Cambridge Univ. Press.

Hartung, J.B., Kunk, M.J., Anderson, R.R. (1990): Global catastrophes in Earth History. Geol. Soc. Am., Spec. Pap. **247**, 207–222.

Hodge, P. (1994): *Meteorite craters and impact structures of the Earth*, Cambridge Univ. Press.

Holsapple, K.A. (1993): The scaling of impact processes in planetary sciences. Annual Rev. Earth Planet. Sci. **21**, 333–373.

Izett, G.A., Cobban, W.A., Obradovich, J.D., Kunk, M.J. (1994): The Manson impact structure: $^{40}Ar/^{39}Ar$ age and its distal impact ejecta in the Pierre Shale in southeastern South Dakota. Science **262**, 729–732.

Izett, G.A., Masaitis, V.L., Shoemaker, E.M., Dalrymple, G.B., Steiner, M.B. (1994): Eocene age of the Kamensk buried crater in Russia. LPI Contr. **825**, 55.

Jansa, L.F., Aubry, M.P., Gradstein, F.M. (1990): Comets and extinctions: Cause and effects? Geol. Soc. Am., Spec. Pap. **247**, 223–232.

Koeberl, C., Sharpton, V., Harrison, T.M., Sandwell, D., Murali, A.V., Burke, K. (1990): The Kara/Ust-Kara twin structure; A large-scale impact event in the late Cretaceous. Geol. Soc. Am., Spec. Pap. **247**, 233–238.

Kolesnikov, E.M., Nazarov, M.A., Badjukov, D.D., Shukulov., Y.A. (1988): The Karskiy craters are the probable record of catastrophe at the Cretaceous-Tertiary boundary. LPI Contr. **673**, 99–100.

Melosh, H.J. (1989): *Impact cratering: A geologic process*, Clarendon.

Menichella, M., Paolicchi, P., Farinella, P. (1996): The Main Belt as a source of Near-Earth Asteroids. Earth, Moon and Planets **72**, 133–149.

Montanari, A. (1991): Authigenesis of impact spheroids in the K/T boundary clay from Italy: new constraints for high resolution stratigraphy of terminal Cretaceous events. Journ. Sed. Petr. **61**, 315–339.

Montanari, A., Asaro, F., Kennett, J.P., Michel, E. (1993): Iridium anomalies of Late Eocene age at Massignano (Italy) and in ODP Site 689B (Maud Rise, Antarctica). Palaios **8**, 420–437.

Montanari, A., Deino, A., Drake, R., Turrin, B.D., De Paolo, D.J., Odin, S.G., Curtis, G.H., Alvarez, W., Bice, D.M. (1988): Radioisotopic dating of the Eocene-Oligocene boundary in the pelagic sequence of the Northrn Apennines. In Premoli Silva I., Coccioni R., Montanari A. (eds.) *The Eocene-Oligocene Boundary in the Marche-Umbria Basin (Italy)*, IUGS Spec. Publ., Aniballi Publ., Ancona, pp. 195–208.

Montanari, A., Deino, A., Coccioni, R., Langenheim, V.E., Capo, R., Monechi, S. (1991): Geochronology, Sr isotope stratigraphy, magneto-stratigraphy and plankton stratigraphy across the Oligocene-Miocene boundary in the Contessa section (Gubbio, Italy). Newsletters on Strat. **23**/3, 151–180.

Montanari, A., Hay, R.L., Alvarez, W., Asaro, F., Michel, H.V., Alvarez, L.W., Smit, J. (1983): Spheroids at the Cretaceous-Tertiary boundary are altered impact droplets of basaltic composition. Geology **11**, 668–671.

Nazarov, M.A., Kolesnikov, E.M., Badjukov, D.D., Masaitis, V.L. (1989): Reconstruction of the Kara impact structure and its relevance to the K/T boundary event. Lunar Planet. Sci. **20**, 766–767.

Pohl, J. (1987): Research in terrestrial impact structures: Introduction. In Pohl J. (ed.) *Research in terrestrial impact structures*, F. Vieweg and Sohn, Wiesbaden, pp. 1–4.

Raup, D.M., Sepkoski, J.J. Jr. (1986): Periodic extinctions of families and genera. Science **231**, 833–836.

Sepkoski, J.J. Jr. (1990): The taxonomic structure of periodic extinction. GSA Spec. Paper **247**, 33–44.

Swisher, C.C. III, Grajales-Nishimura, J.M., Montanari, A., Margolis, S.V., Claeys, P., Alvarez, W., Renne, P., Cedillo-Pardo, E., Maurasse, F., Curtis, G.H., Smit, J., McWilliams, M.O. (1992): Coeval $^{40}Ar/^{39}Ar$ ages of 65.0 million years ago from Chicxulub Crater melt rock and Cretaceous-Tertiary boundary tektites. Science **257**, 954–958.

Earth-Orbiting Debris Cloud and Its Collisional Evolution

Alessandro Rossi[1] and Paolo Farinella[2,3]

[1] Istituto CNUCE del CNR, Via Santa Maria 36, I-56126 Pisa, Italia
[2] Dipartimento di Matematica, Università di Pisa, Via Buonarroti 2, I-56127 Pisa, Italia
[3] O.C.A. Observatoire de la Côte d'Azur, Dept. Cassini, B.P. 4229, F-06304 Nice Cedex 4, France

Evolution collisionnelle dans le nuage de débris spatiaux autour de la Terre

Résumé. Plus de 8000 objets de dimension supérieure à 20 cm orbitant autour de la Terre sont actuellement répertoriés et surveillés par le "US Space Command". Seulement 350 d'entre eux sont des satellites en activité. Des études récentes indiquent que le nombre d'objets encore plus petits (mais supérieurs à 1 cm) se situe entre 130 000 et 260 000, et que ceux supérieurs à 1 mm pourrait être plus de 36 millions.
Le danger potentiel que représente la prolifération croissante de ces "débris orbitaux" est à l'origine des études de suivi de l'évolution à long terme de cette population. Les mécanismes de formation et de disparition à prendre en compte sont : les lancements spatiaux et les explosions qui apportent de nouveaux matériaux; la descente des objets sur des orbites plus basses à cause du frottement atmosphérique; et, finalement, l'influence des collisions mutuelles sur la distribution en taille de ces objets par génération d'essaims de fragments. Un paramètre fondamental pour simuler l'évolution collisionnelle de la population des débris est, bien sûr, la probabilité de collision entre ces objets. Nous avons calculé les probabilités de collision à chaque altitude pour 2700 objets orbitaux réels.
Nous avons aussi développé un algorithme numérique pour modéliser l'évolution collisionnelle future de la population de débris en orbite terrestre basse, en tenant compte à la fois du large spectre de masse (selon les résultats expérimentaux disponibles) des débris orbitaux, et de leurs différentes altitudes qui résultent de l'efficacité variable du freinage atmosphérique. Dans ce modèle, tous les cas montrent une croissance fortement exponentielle du nombre des fragments : on prédit que cette croissance s'emballera d'ici le prochain siècle, d'abord dans la couche très peuplée entre 700 et 1000 km d'altitude puis, peu après, entre 1400 et 1500 km. Cet emballement est retardé seulement de quelques siècles si le seuil de fragmentation catastrophique pour l'énergie spécifique d'impact des objets orbitaux dépasse ceux des corps rocheux naturels d'au moins un facteur 10. Nos simulations montrent que la sensitivité des résultats par rapport aux lancements futurs et/ou aux politiques de "nettoyage" est très faible, de telle sorte que des mesures énergiques doivent être prises bientot si nous voulons éviter la catastrophe.

Abstract. The growing hazard related to the proliferation of Earth-orbiting debris is leading to the study of realistic models for the long-term evolution of this population.

Few source and sink mechanisms affect the evolution: launches and explosions provide new material, partially subject to human control, the object decay to lower shells due to the atmospheric drag (according to their altitude and area to mass ratio) and finally the orbiting objects can undergo high-velocity mutual collisions which affect their size distribution by generating swarms of fragments, playing the role of potential new projectiles. In order to simulate the collisional evolution of the debris population, we calculated the actual intrinsic collision probabilities for the orbiting objects for each altitude. We also developped a numerical algorithm to model the future collisional evolution of the low-orbiting Earth debris population. Our simulations show that the sensitivity of the results to future launch and/or deorbiting and removal policy is rather weak, so that drastic measures will nead to be taken soon in order to significantly avoid or delay a catastrophic outcome.

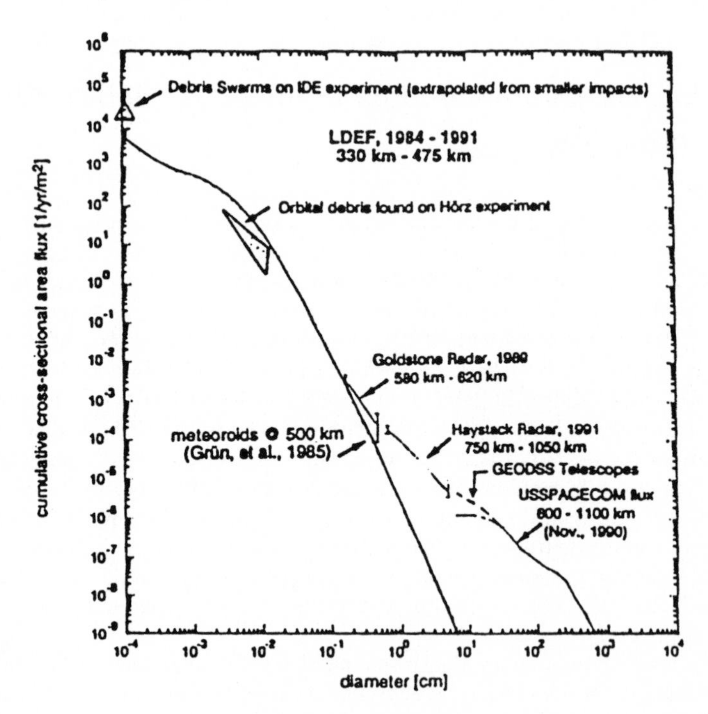

Fig. 1. Meteoroid environment compared to recent measurements of orbital debris environment.

1 Introduction

- October 4, 1957: Beginning of the space age, the USSR orbits Sputnik 1, the first artificial satellite.
- June 29, 1961: first known satellite break-up.
- July 24, 1983: the Space Shuttle Challenger (STS mission 7) lands at the Edwards Air Force Base, California. In one windshield there is a small "crater".

The window must be replaced before the next flight. It has been hit by a fleck of paint no larger than 0.2 mm in diameter. Hundreds of millions of such particles are probably in orbit around the Earth.

- 1988: The Space Debris Working Group, set up by the European Space Agency, concludes its report by stating that: *"The self-sustained debris production by collisions is a long-term concern. It is however the most far-reaching threat which could terminate all space activities. This mechanism requires further careful study."*
- February, 1994: At the AIAA Conference (Cocoa Beach, Florida), Robert Crippen, pilot of the first Space Shuttle and currently Director of Kennedy Space Center (KSC), Shuttle Launching Base, answers the question *"Which astrodynamical problem do you think will be the most important to be solved for the realization of the Space Station?"* by replying simply: *"Space debris"*.

We will describe in the next sections what the space debris problem is, where it does come from, which are the forecasts for the future and which measures have been proposed to solve the problem.

2 The Near-Earth Environment and the Space Surveillance

From the mm-sized particles up to the largest satellites, tens of meters across, the population of objects of artificial origin orbiting in near-Earth space spans many orders of magnitude in size and mass. Below about one millimeter in diameter the natural meteoroid population still dominates the particle flux, whereas for larger sizes artificial space debris accounts for the majority of the solid bodies present in the near-Earth environment (Fig. 1).

In near-Earth space there are two major regions where orbital debris is of concern: Low Earth Orbits (LEOs), below about 2000 km, and Geosynchronous Orbits (GEOs), at an altitude of about 36000 km. The issues are in principle the same in the two regions, nevertheless they require different approaches and solutions. In this paper our discussion will be focused mainly on LEOs, where the situation appears to be deteriorating faster.

About 5000 Earth-orbiting payloads have been launched since the first satellite in 1957. Among these objects, some 2500 have re-entered in the atmosphere. The others (about 2500) are still orbiting, including 10 and 33 satellites carrying on board radioisotope thermoelectric generators and deactivated nuclear reactors, respectively. It is important to note that in the population of orbiting payloads, only about 20% (i.e. some 500 objects) are active satellites and probes. The launch rate is presently about 80 per year. It reached its peak at the end of the 60s and the mid-70s (e.g., 135 launches in 1967 and 132 in 1976) and has now decreased, mainly due to economical and political reasons (in particular those concerning the former Soviet Union, the largest spacefaring country). It should also be pointed out that only a small fraction of the spacecraft being launched (e.g., the NASA Space Shuttles and many Russian reconnaissance satellites) re-enter under active control at the end of their operational lifetime.

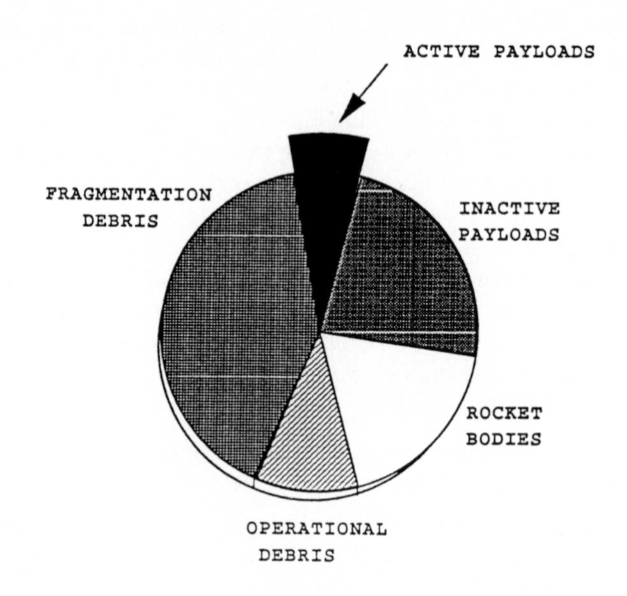

Fig. 2. Make-up of the USSC-catalogued Earth satellite population.

All the satellites currently in orbit are catalogued by the United States Space Command (USSC) in the so called Two-Line Orbital Element Catalogue. In this catalogue about 8300 objects are listed, along with their current orbital parameters. Thus, besides the payloads, it contains also spent rocket bodies and pieces of debris (about 16% and 58% of the total, respectively; see Fig. 2).

The latter category is composed partly by "operational debris" (e.g., a dozen screws missing from a 1984 Shuttle flight, a screwdriver which floated away from a cosmonaut space-walking outside the *Mir* station, camera lens caps, yo-yo masses, and so on), but mostly it includes the products of 142 fragmentation events, which have involved rocket upper stages or spacecraft in orbit. 42 of these events had unknown causes, 50 were deliberate (to avoid the possible retrieval of spy satellites by other countries, to test anti-satellites weapons, to make structural tests), 41 were propulsion-related and 9 were due to electrical or command failures (see Fig. 3). Although not certain, it is believed that at least 2 of the events having unknown causes have been probably due to unintentional collision-induced break-up events.

The objects listed in the USSC catalogue, with an overall mass of about 3,400 metric tons, accounts for about 99.9% of the mass in orbit. However, the catalogue is supposed to be 90% to 99% complete only for diameters larger than 20 cm in LEOs (and 1 meter in GEO), whereas objects smaller than 10 cm are almost absent from the catalogue (only some 10% to 25% are included in the range from 1 to 10 cm).

For the detection and measurement of the population of space objects two methods have been used so far: ground-based observations, by radar and optical techniques, and in-flight data. The routine activity of detection, tracking and

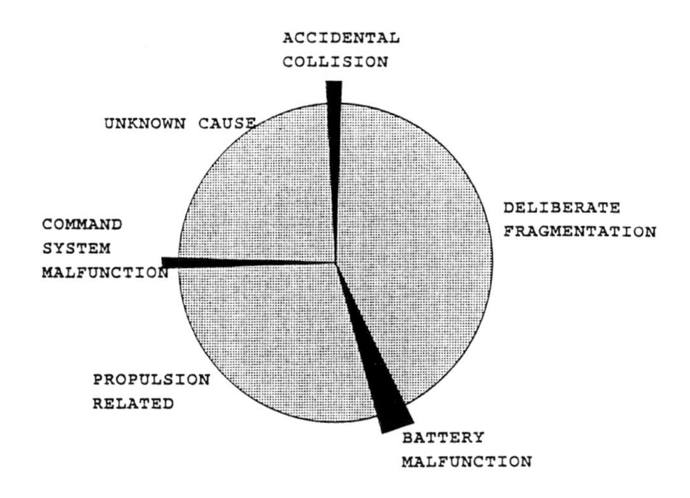

Fig. 3. Relative proportions of satellite break-up events by cause.

cataloguing is performed mainly by the Space Surveillance Network (SSN), coordinated by the Space Surveillance Center (SSC) located at Cheyenne Mountain, Colorado. The SSN is composed by 26 sensors widely disperded in the US territory. The radars include 9 mechanically steered dishes, one radar interferometer (the NAVSPASUR "radar fence") and 9 large phased-array radars capable of tracking several objects simultaneously.

Above 5000 km of altitude the radar power is not enough to monitor the small space debris, as the returned flux is proportional to the -4 power of the distance, so the SSN uses optical sensors for the higher objects. Until 1987 the well-known Baker-Nunn photographic systems were the primary sources for optical informations; they had a limiting magnitude of 14 and used photographic plates, which required long processing. Therefore, after 1987 they have been replaced by electro-optical devices, which compose the Ground-based Electro-Optical Deep Space Surveillance System (GEODSS), including 6 different observing sites. This system achieves accurate pointing and a very good sensitivity (limiting magnitude 16.5). Tracking and data processing is now automated. The last sensor of the SSN is a passive radio-frequency sensor (PASS). Some 40,000 to 70,000 observations are processed daily by the SSC.

To get data on the cm- and mm-sized particles different sensors, or the same sensors but operated in a different way, are needed. Experimental radar and optical campaigns have been carried out to detect 1 cm objects from the ground. The optical campaigns performed in recent years have reached a limiting size of about 8 to 10 cm, and brought to the conclusion that the USSC Catalog is only about 50 % complete at about 10 cm, in LEOs. To detect 1-cm objects in LEO it was necessary to exploit radar sensors. The technique used is to put the radar in a "beam park" mode, where the radar stares in a fixed (usually vertical) direction and the debris randomly passes through the field of view and is detected. This

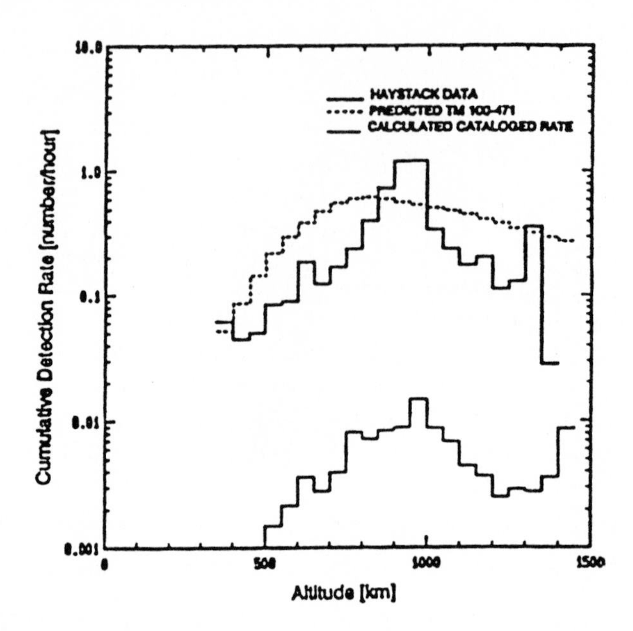

Fig. 4. Rate of detection by the Haystack Radar vertically pointed for 188 hours of data.

allows only counting of the number of objects but not a determination of their orbits. A large-scale campaign with this technique has been performed recently with the Haystack radar. Several hundreds of hours of data have been collected and analyzed; at the lowest altitudes (350 km) objects down to 3 mm were detected and at the highest altitude (1400 km) the limit was just 6 mm. The detection rate (Fig. 4) averaged over all altitudes is about 65 times the rate predicted by the catalogue alone; in particular in the altitude band between 850 and 1000 km (where a large number of break-up events happened) the rate is 100 times that predicted by the catalogue.

Anyway, all the above-mentioned sensors have a basic limitation, since they are mainly concentrated at high latitudes in the Northern hemisphere. So they are not able to appropriately monitor the very low inclination orbits (especially the highly elliptical Geosynchronous Transfer Orbits, GTOs) and the high eccentricity/high inclination orbits with perigee in the Southern hemisphere (e.g., the so-called Molniya orbits).

This problem has been highlighted by the analysis of the surfaces of the Long Duration Exposure Facility (LDEF). LDEF is a satellite released by the Space Shuttle in 1985 and returned to Earth in 1991. It is the best source of in-flight data on space debris and meteoroids to date (other sources are the shuttle windows and few other satellites returned to ground by the Shuttle after their operational life). LDEF was gravity-gradient stabilized and maintained a fixed orientation with respect to the Earth. So the analysis of its surfaces provides information also on the directions of the incoming projectiles. 14% of the impacts on the front and side surfaces were identified as due to orbital debris; 55% of the impacts were of inidentified origin but most of them are supposed to come from man-made debris either. A surprising result, which revealed a gap in the ground measurements, was that 15% of the impacts on the rear-facing surfaces (with respect to the orbital motion) were related to orbital debris. This implies that a large amount of undetected debris lies in elliptical orbits (the only ones capable of hitting LDEF from rear), probably coming from fragmentations of orbital transfer stages, both US ones, with inclinations between 25° and 28.5° and ESA ones, with inclinations between 4° and 10°. The relative number of the catalogued objects of this kind must be increased by a factor 20 to be consistent with LDEF measurements. If all inclinations are taken into account, the number of objects in highly elliptical orbits should be increased by a factor 30.

The current estimate is that the total number of non-trackable particles of 1 cm and greater is between 130,000 and 260,000, while those larger than 1 mm could be more than 36,000,000.

In Figs. 5, 6 and 7 the orbital ditribution of all the objects listed in the USSC catalogue is plotted.

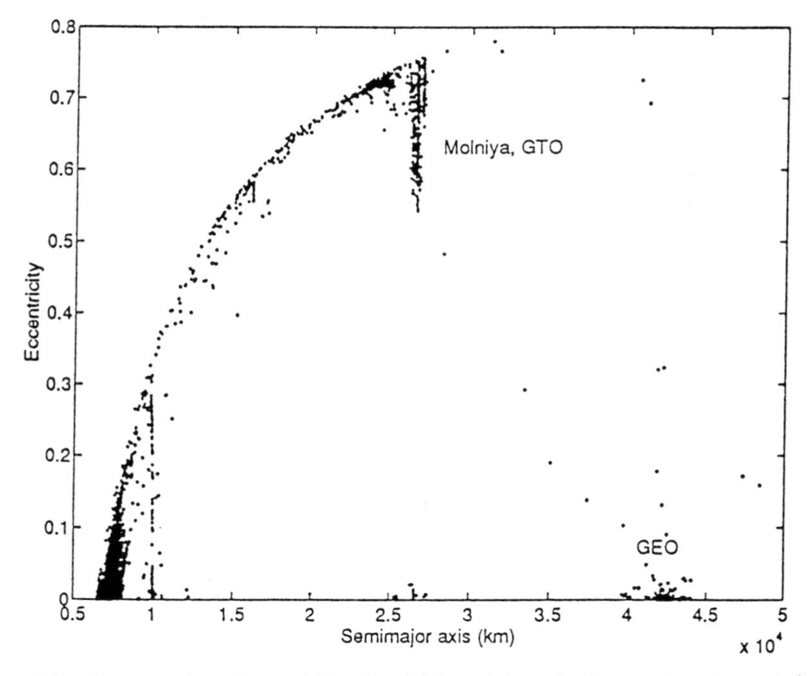

Fig. 5. Distribution of catalogued Earth-orbiting objects in the semi-major-axis/eccentricity plane.

They are clearly grouped in "families" or constellations, according to their different purposes and to the different launching bases: we can find out the US GPS (Global Positioning System) satellites and their Russian analogues GLONASS ($a \simeq 26,000$ km, $i \simeq 55°$ and $i \simeq 63°$, respectively), the Russian communication satellites in Molniya-type orbits ($a \simeq 26,000$ km, $e \simeq 0.7$, $i \simeq 63°$), the geosynchronous satellites ($a \simeq 42,000$ km, $e \simeq i \simeq 0$), the satellites in Sun-synchronous orbits ($i \simeq 100°$) , the satellites in polar orbits ($i \simeq 90°$), some families of russian COSMOS satellites between $i \simeq 60°$ and $i \simeq 80°$, the LEO satellites launched from the Kennedy Space Center (at $i \simeq 27°$) and the families of objects in geosynchronous transfer orbits (mostly upper stages) launched from Kourou (ESA Ariane rockets, $i \simeq 7°$), from the Kennedy Space Center ($i \simeq 27°$) and from Baikonour ($i \simeq 48°$).

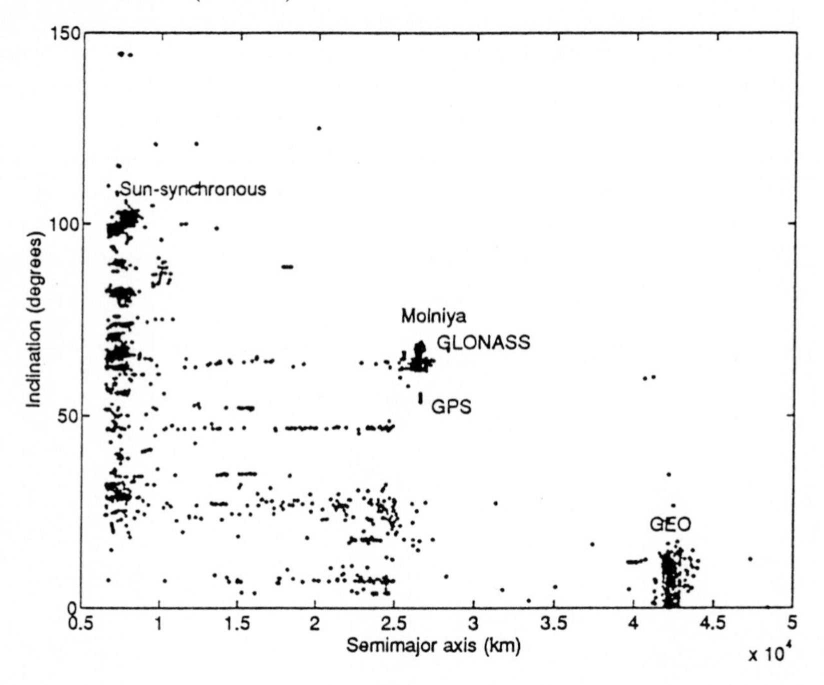

Fig. 6. Distribution of catalogued Earth-orbiting objects in the semi-major-axis/inclination plane.

3 Long-Term Evolution of the Debris Population

As a matter of fact, the low-orbiting Earth debris population is similar to the asteroid belt, since it is subject to a process of high-velocity mutual collisions that affects the long-term evolution of its size distribution. However, the situation is more complex than for the asteroids, because here the source and sink mechanisms are (partially) subject to human control (e.g. launches, explosions and retrievals), the number density of objects is a sensitive function of the altitude (and so is the sink mechanism due to drag) and the relative speeds are dominated by mutual inclinations, which are much larger than typical orbital eccentricities and unevenly distributed (whereas among the asteroids eccentricities and inclinations have similar, fairly broad distributions, with average values

≈ 0.15). We shall now review the main processes contributing to the long-term evolution of the Earth-orbiting population.

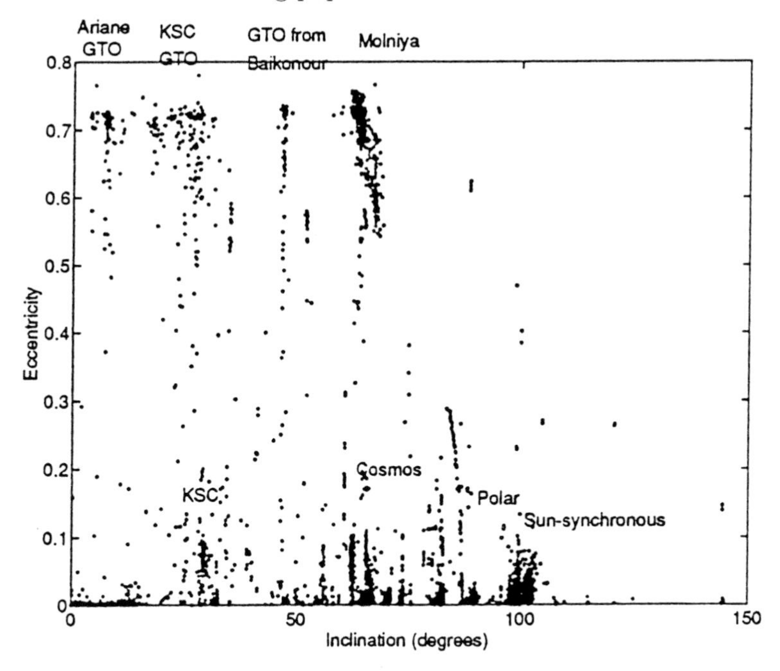

Fig. 7. Distribution of catalogued Earth-orbiting objects in the eccentricity / inclination plane.

3.1 Source and Sink Mechanisms

Launches. The main source of new large objects is represented by launches. As we mentioned earlier, the average number of launches is about 90-100 per year, performed mostly by Russia and the USA. It is very difficult to predict the future trend of the launching activity, since it is strictly related to technological, economical and political problems. Anyway, mainly due to the crisis in the former Soviet Union, a decrease in the number of launches has already taken place and can be foreseen to continue for the next years. With every launch, an average number of 1.2 payloads has been injected into orbit in the last decade. Besides the payload, that is of course the primary objective of the launch, some "junk" is usually left in space. This includes small objects like the yo-yo masses used to slow down the rotation of the satellite and the covers of the satellite sensors, but also large upper stages, i.e. the engines used to inject the spacecraft in its final orbit. The upper stages represent a significant danger, since sometimes they contain residual fuel which can lead to an explosion, generating a large amount of small fragments.

Explosions. The in-orbit explosions are responsible for most of the current debris population. Usually the explosions can be classified into two different types: low-intensity and high-intensity ones. A low-intensity explosion can be defined as an event caused by a charge not being in direct contact with the spacecraft structure, such as a pressure vessel explosion. A high-intensity explosion instead is assumed to occur when a charge is in direct contact with the spacecraft structure, e.g. due to the ignition of excess fuel. *Ad hoc* experiments have been performed to obtain data for high-intensity explosions, whereas the fragments from an exploded Atlas missile were used to derive experimental results for low-intensity events. The mass distribution for both types of explosions approximately match exponential laws of the form (Bess, 1975):

$$N(>m) = N_0 e^{-c\sqrt{m}} ,$$

where $N(>m)$ is the number of fragments with mass larger than m, c a constant to control the slope of the curve and N_0 the total number of fragments. High-intensity explosions have tipically larger values of both c and N_0 than low-intensity ones. Different choices of these parameters have led to many relations that best fit the outcomes of different experiments in various mass ranges. In particular, in our long-term evolution model (to be described later), we adopted the relationship proposed by Su and Kessler (1985):

$$N(m) = \begin{cases} 1.71 \times 10^{-4} M_t e^{-0.02056\sqrt{m}} & \text{for} m \geq 1936\, g \\ 8.69 \times 10^{-4} M_t e^{-0.05756\sqrt{m}} & \text{for} m < 1936\, g \end{cases}$$

M_t being the mass of the exploded spacecraft in grams.

Collisions: Impact Rates and Velocities. Even if a disruptive collision between two artificial objects in space has not yet been reported with certainty, it is clear that in the future hypervelocity impacts may become the driving mechanism of a possible catastrophic growth of the orbiting debris population.

An essential tool needed for the study of the long-term evolution of space debris is a quantitative estimate of the frequency of collisions in Earth orbit and the characteristic impact speed. For a given target object, both quantities depend on the distribution of orbital elements in the assumed projectile population, and can be estimated as follows.

The number of collisions between two orbiting particles expected during a time interval Δt can be formally expressed as $P_i\ (R+r)^2\ \Delta t$, where r and R are the projectile and target radius, respectively (both bodies are assumed to be spherical here for the sake of simplicity, but for other shapes suitable "mean radii" can be easily defined), and P_i is the so-called *intrinsic collision probability*, a quantity depending only on the two sets of orbital elements; P_i may be interpreted as the collision frequency between two bodies for which $(r+R) = 1$ meter. Of course, $P_i = 0$ if the two orbits cannot intersect each other for geometrical reasons; this occurs when the apogee distance of the inner orbit is smaller than the perigee distance of the outer one.

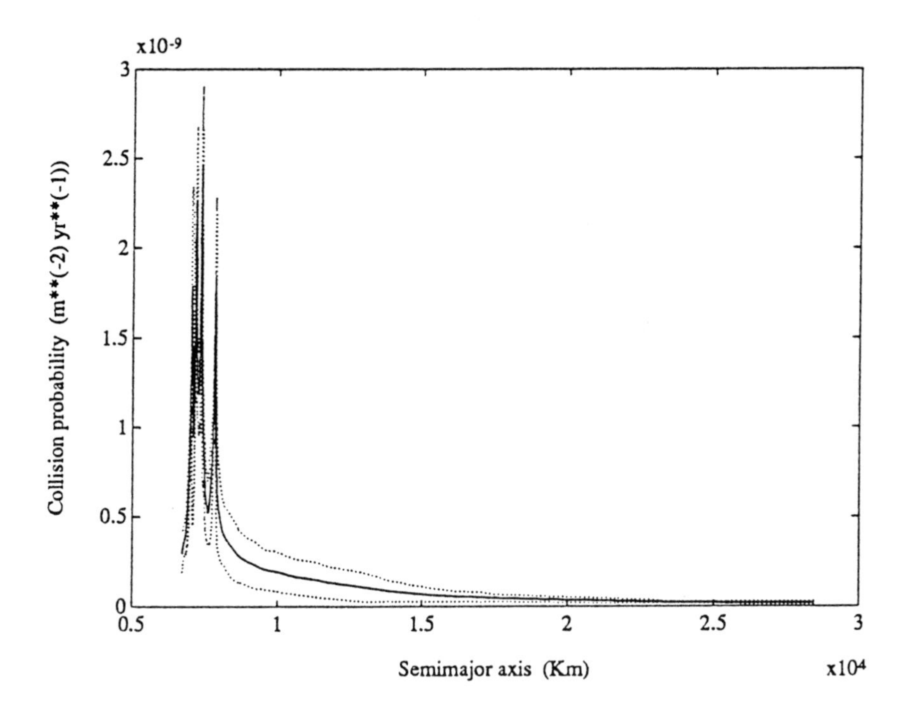

Fig. 8. Running-box mean of the collision probability ($m^{-2}yr^{-1}$) with $\pm 1\sigma$ variation, as a function of the semi-major axis.

A simple, approximate estimate of P_i can be obtained by applying a particle-in-a-box (PIAB) model, that is by assuming that all bodies are homogeneously distributed in an interaction volume, W, where they can move freely between collisions with their own random speeds. For any pair of bodies, the average number of collisions in a time interval Δt can be easily estimated as $AV\Delta t/W$, where V is the average relative velocity and A is the collision cross section. Therefore, we get $P_i \approx \pi V/W$. For the LEO region (say, between 200 and 2000 km altitude), we have $V \approx 10$ km/s and $W \approx 10^{12}$ km^3, and therefore the PIAB model yields a mean intrinsic collision probability of $\approx 10^{-9}$ m^{-2} yr^{-1}.

Of course, the basic assumptions of the PIAB model are incompletely satisfied for a set of orbiting bodies, because for them there is no well-defined box volume and, under the effects of the primary's gravity and different perturbations, the relative motion is not completely random. To derive intrinsic collision probabilities for the LEO population we applied the algorithm developed by Wetherill (1967) to a set of 2700 actual orbiting objects (Rossi and Farinella 1992). Since the computational task was very heavy we parallelised a sequential application code used by Farinella and Davis (1992) for the asteroid case and used a *nCUBE 2* multicomputer, whose architecture is composed of 128 independent CPU nodes.

Due to the typically small eccentricities, every object on the average crossed only 43.4% of the other orbits; actually this results from a majority of orbits with less than 1200/2697 crossings, and a minor peak of high-eccentricity orbits which can cross almost all the other

nes. The distribution of the average intrinsic collision probability is fairly broad: some 10% of the orbits yield very small values, while the remaining ones are evenly distributed up to $\approx 3 \times 10^{-9}$ m^{-2} yr^{-1}. The objects having small values of P_i (typically, a few percent of the average) are mostly the high-eccentricity ones, with semi-major axis $> 20,000$ km, which spend only a small fraction of their orbital period in the crowded LEO shell. This distribution of P_i yields an overall mean and standard deviation of $1.105 \pm 0.812 \times 10^{-9}$ m^{-2} yr^{-1}. Note that this average value is very close to the PIAB prediction, although values a factor 2 lower or higher than the average are also common. On the contrary, the distribution of the average impact velocities is fairly narrow, and yields an overall mean and standard deviation of 9.654 ± 0.875 km/s, fairly close to the commonly quoted value of 10 km/s. We have produced running-box plots of the average intrinsic collision probability and impact velocity as a function of the orbital elements (Figs. 8 and 9); the average value of P_i is a fairly sensitive function of the orbital elements and Fig. 8 displays clear peaks at the same semi-major axes where many objects are clustered.

It is worth stressing that the average value of P_i given above means that a space station of 50 m^2 cross-section in LEO has a typical collision rate of about 0.09% yr^{-1} with the approximately 50,000 orbiting objects larger than 1 cm, which are generally considered as capable of producing catastrophic damage to operational spacecraft. Encounters within a closest approach distance < 1 *km* with the same potential projectiles occur about once per week, implying the need for frequent active avoidance manoeuvres. These estimates have an uncertainty of at least a factor two owing to the poorly known abundance of the projectile population; moreover, the distribution of P_i is such that, for a fixed projectile population (provided the distribution of orbital elements is similar to that of our sample) the actual collision/encounter rate can grow by up to a factor 3 times the average value depending on the chosen orbital elements.

Collisions: Models for the Collisional Outcomes. Hypervelocity impact experiments carried out in the laboratory with solid "rocky" targets of different compositions and aimed at simulating interasteroidal collisions (Fujiwara et al., 1989) have shown that the cratering and break-up regimes are associated to different typical values of the collisional parameter given by the ratio between the projectile kinetic energy (E_p) and the target mass (M_t). Thus, collisional outcomes can be modelled assuming that each collision provokes either localized target damage (with a fraction of target mass M fragmented and ejected from a "crater" of mass proportional to the impact energy, up to a maximum of $M/10$), whenever E_p/M_t is lower than a given threshold value Q^* (which can be called *impact strength*, and is shown by experiments to be of the order of 10^3 - 10^5 J/kg); or complete target break-up, with a largest fragment including

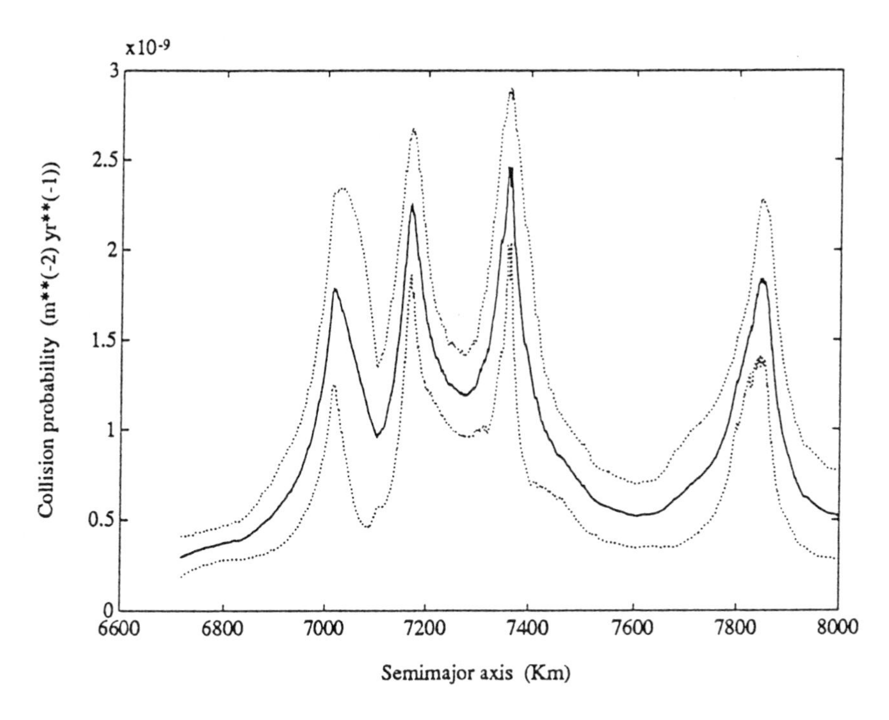

Fig. 9. Detail of the running-box mean of Fig. 8 for the objects with semi-major axis < 8000 km.

less than one half of the target mass, if $E_p/M_t > Q^*$. In such a model, one can also assume that the (approximately constant) crater mass over impact energy coefficient is $(10Q^*)^{-1}$ and that, for a fixed impact velocity V, the threshold projectile-to-target mass ratio for target break-up is just $2Q^*/V^2$. Since this is much less than unity, the mass of the projectile can be neglected in computing the resulting fragment mass distribution.

A number of *ad hoc* collision experiments with mock-up spacecraft bodies have been carried out in order to derive empirical analytical expressions for the mass and velocity distributions of the fragments. Unfortunately, most of these experiments have been performed in military installations and the results are classified. In many cases analytical relationships can be fitted to the data, but they strongly depend on the geometry and composition of the target and the projectile. A report by the Physical Sciences Incorporated (Nebolsine et al. 1983) has analyzed data from hypervelocity collision tests conducted at the Arnold Engineering and Development Center. The analysis showed two distinct portions of the mass distribution for a collision-induced spacecraft fragmentation. At very small fragments sizes (ejecta that are directly produced by the impact in the vicinity of the collision site), a power-law distribution was observed. On the other hand, the larger pieces follow an exponential law of the same form as that associated with explosions. This can be interpreted in the sense that the large pieces are just an indirect result of the collision, deriving from the break-up of the remainder of the satellite. The transition from the power law to the exponential law occurs at about 1 cm in diameter.

In the long-term evolution model described later on, we adopted an even simpler semi-empirical model, namely we assumed that in both the cratering and the break-up cases, the fragment mass distribution can be represented by a truncated power law, up to the mass of the largest fragment (Petit and Farinella 1993). In the cratering case, the characteristic exponent q of the incremental fragment mass distribution $dN \propto m^{-q}dm$ (where dN is the number of fragments having masses in the interval $[m, m + dm]$) has been fixed to 1.8; mass conservation implies that in this case the largest fragment comprises 1/4 of the total excavated mass. In the break-up case, the following empirical relationship between the largest fragment to target mass ratio m_1/M and the specific impact energy E/M normalized to Q^* has been used:

$$\frac{m_1}{M} = \frac{1}{2}\left(\frac{E}{Q^* M}\right)^{-1.24}$$

(Fujiwara et al. 1977), whereas q can be derived from the mass conservation relationship and turns out to be

$$q = \frac{(2 + m_1/M)}{(1 + m_1/M)},$$

implying that it ranges from 5/3 in a barely catastrophic collision ($m_1/M = 1/2$) to 2 in a supercatastrophic event ($m_1/M \to 0$), the latter case being such that equal logarithmic mass bins contain equal masses.

Drag and Retrievals. For many LEO satellites atmospheric drag changes the orbit in a significant way. The corresponding perturbative forces dissipates the satellite's orbital energy, causing the semi-major axis to decrease in a secular way. This spiral-like motion eventually causes a sudden re-entry of the satellite onto the Earth's surface, and this is the only "natural" mechanism which can remove objects from space.

The drag force F_D is approximately proportional to: (i) the satellite's cross-section normal to the atmospheric flow; (ii) the atmospheric density; and (iii) the square of the satellite's velocity relative to the atmosphere. The corresponding acceleration is of course inversely proportional to the satellite's mass. When the orbit shrinks, both the density and the orbital speed increase, causing the orbit decay to further speed up, and this positive feedback mechanism explains why the final re-entry occurs in a "catastrophic" fashion, over a small portion of a revolution, with little or no possibility of predicting the corresponding location and timing in reliable way. Also, since the atmospheric density decreases with altitude in a quasi-exponential way, and has diurnal and long-period variations (the latter due to the periodically changing solar activity), small, low-orbiting particles are more effectively removed by drag, and this "cleaning" mechanism is faster near the peaks of the solar cycle. As for meter-sized satellites, atmospheric drag is effective as a removal mechanism only for very low orbits (more exactly, for low values of the perigee distance), typical lifetimes increasing from ≈ 1 to 10^3 yr between 400 and 1000 km altitudes. These lifetimes are inversely proportional to the area-to-mass ratio of the orbiting object.

We recall here that, since the space debris problem has become salient, a number of strategies has been proposed to remove actively the"dead" satellites from orbit at the end of their operative life (this is normally done only for some particular categories of spacecraft). The proposed measures range from the compulsory re-entry of the spent upper stages in the atmosphere to the adoption of new devices to actively go and retrieve spent satellites from crowded orbits. We will come back to these proposals in Sects. 3.2 and 4.

3.2 Long-Term Evolution Model

We will describe here a numerical model to simulate in a semi-deterministic fashion the future evolution of the debris population, that we have developed in recent years in collaboration with L. Anselmo, A. Cordelli and C. Pardini (Cordelli et al. 1993; Rossi et al. 1994).

The model requires the numerical integration of a set of 150 coupled, non-linear, first-order differential equations, with each equation giving the rate of change of the population present in a discrete size bin and in a given altitude shell. We have used 15 altitude shells (six 50-km thick shells between 400 and 700 km, plus nine 100-km thick ones up to 1600 km) and 10 logarithmic mass bins (centered at values ranging from 1 g to 6000 kg, and spanning a factor 5.664 each).

Initial conditions are provided from the (limited) knowledge of the existing population, which has to be extrapolated to the smaller size range of untrackable particles. The assumed initial population consists of about 57,000 bodies, all exceeding a minimum mass of 0.42 g: the vast majority of these objects have masses smaller than a few tens of grams, and only about 1650 exceed 10 kg. We simulated all the source and sink mechanisms described in Sec. 3.1 (in particular, two explosions per year are assumed to occur on average in each of the two altitude shells between 700 and 900 km, involving bodies of mass $M_t = 1500$ kg). The net current rate of insertion into orbit of new massive objects is assumed to be about 60/year overall, the impact strength Q^* adopted is 10^3 J/kg, similar to that found experimentally for natural stony targets.

The corresponding evolution equations for $N(m_i, h_j, t)$, the number of objects residing at time t in the bin centered at mass m_j and in the shell centered at height h_j, read:

$$\frac{dN(m_i,h_j,t)}{dt} = \beta(m_i,h_j) - \frac{N(m_i,h_j)}{\tau(m_i,h_j)} + \frac{N(m_i,h_{j+1})}{\tau(m_i,h_{j+1})} + \sum_{k,l} f(m_k,m_l,m_i)P(h_j)\sigma(m_k,m_l)N(m_k,h_j)N(m_l,h_j)$$

where β is the matrix that accounts for the launches, the release of small "operational debris" and the explosions, as specified above, whereas the second and third terms account for the drag-induced orbit decay, assuming a characteristic residence time τ depending on both altitude and mass, and the fourth one accounts for the collisions. Here the three-dimensional array $f(m_k, m_l, m_i)$ represents the number of objects of mass m_i produced (or destroyed) by a collision between two bodies of mass m_k and m_l; $P(h_j)$ is the intrinsic collision probability for objects residing in shell j; and $\sigma(m_k, m_l)$ is the squared sum of the radii of the same two bodies, derived from their cross-sections.

Of course many areas of uncertainty remain in this modelling work. The most critical ones appear to be: (i) the relationship between mass and cross-section of the orbiting objects; (ii) the value of the (average) impact strength Q^* of the targets and the fragment mass distribution function, which may depend on the shape, structure and material properties of the typical spacecraft; (iii) the existing Earth-orbiting populations as a function of mass and altitude, which are the initial conditions for the future evolution; (iv) the future launch and explosion rates, the two dominant sources of new objects before disruptive impacts become frequent enough; (v) the time dependence (neglected in our model) of drag in the high atmosphere, which is mainly associated with the 11-yr solar activity cycle and may affect in a significant way the decay rate for small, low-orbiting bodies; (vi) the effects due to objects in elliptical orbits, in particular fragments generated by high-intensity explosions. Anyway, we selected some nominal values for all the "free parameters" in the model and then performed a sensitivity analysis (as reported in Rossi et al., 1994). In each case our set of 150 differential equations was integrated numerically for a time span of several centuries in the future.

The qualitative evolution pattern in our nominal case is plotted in Figs. 10, 11 and 12, which refer to the second ($m \approx 6$ g), fifth ($m \approx 1$ kg) and ninth ($m \approx 1$ ton) mass bins, respectively. After a period of slow and steady population growth ranging from decades to centuries, depending on the altitude, the generation of collisional fragments exceeds the insertion into orbit of non-collisional debris and significantly increases the frequency of catastrophic impacts. Since each break-up event generates a swarm of new potential "projectiles", thus increasing the subsequent rate of catastrophic events, a kind of chain reaction is triggered. As a consequence, the growth of the small-size population becomes exponential, whereas the abundance of larger objects (including the operational satellites) reaches a maximum and then rapidly drops. In this phase, the environment is dominated by collisional fragmentation, with more satellites being destroyed than launched. At the end, a quasi-steady state is reached, with all the material being launched rapidly converted into fragments. Of course, it is plausible to infer that before this phase is reached, any space activity will have ceased of being carried out in the corresponding altitude shells. The most critical altitude range for the early onset of runaway fragment growth corresponds to the crowded shells between 700 and 1000 km - here the runaway growth of fragments starts about 50 yr in the future. But not much later the process is triggered also between 1400 and 1500 km. Only between 1000 and 1400 km and beyond 1500 km the fragment growth is delayed until 100 to 300 yr in the future.

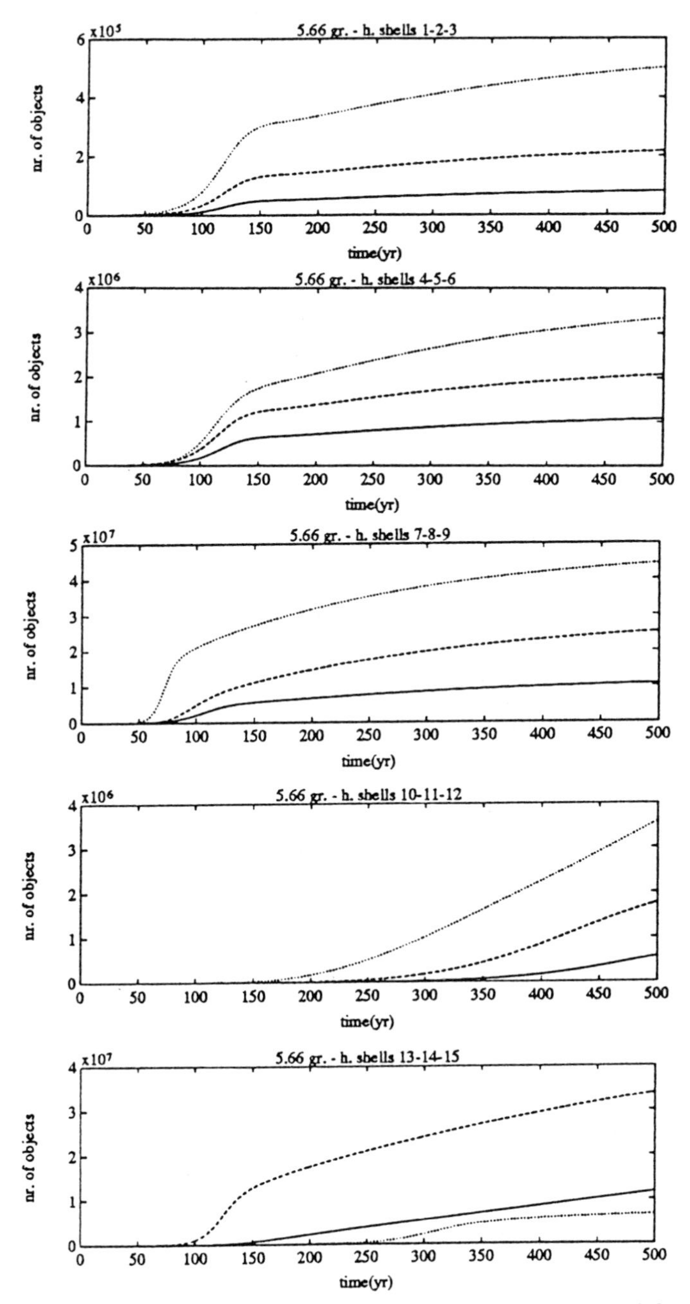

Fig. 10. Nominal case for the time evolution over 500 yr in the future of the populations of orbiting objects residing in the mass bin centered at 5.66 g (and ranging between 2.38 and 13.5 g), for each of the 15 altitude shells considered in our model. In every graph, three shells are plotted together, with the lower, intermediate and higher ones being represented by the solid, dashed and dotted curves, respectively. We recall that the shells are centered at 425 (no. 1), 475 (2), 525 (3), 575 (4), 625 (5), 675 (6), 750 (7), 850 (8), 950 (9), 1050 (10), 1150 (11), 1250 (12), 1350 (13), 1450 (14) and 1550 (15) km of altitude.

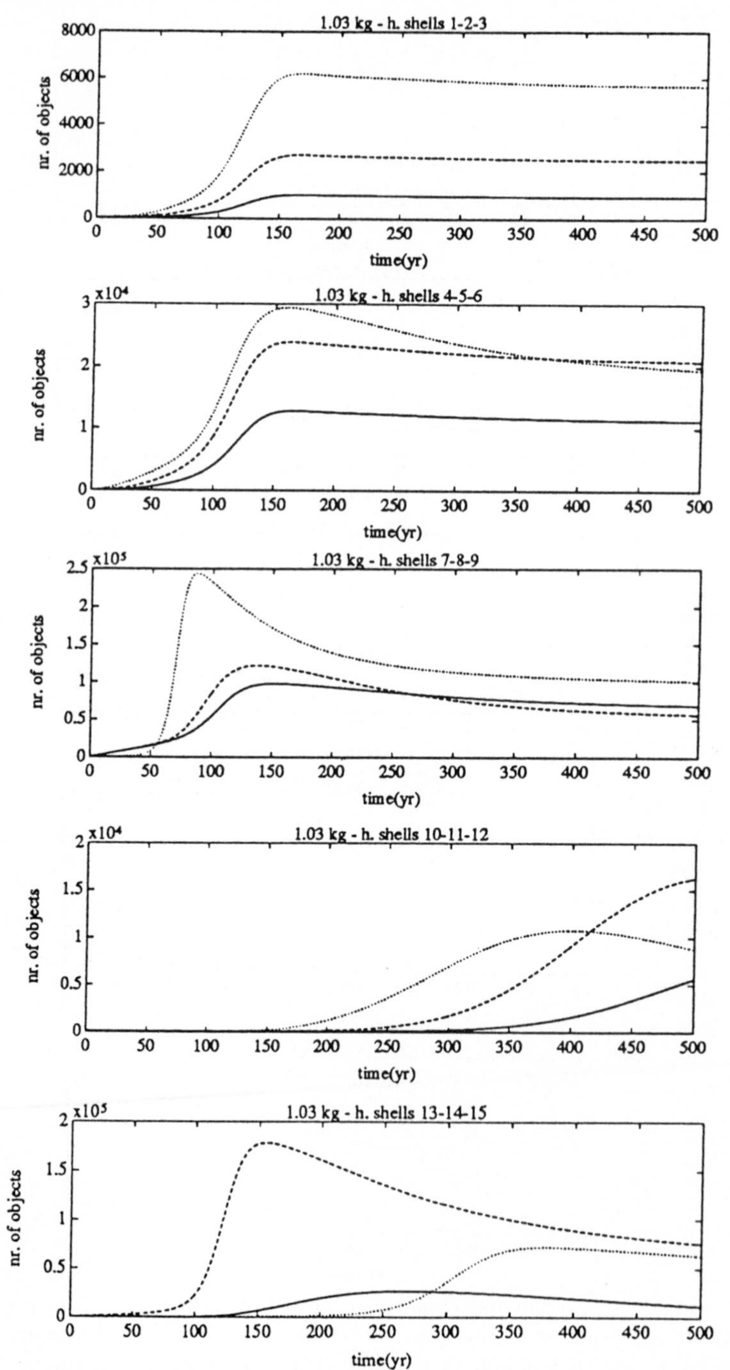

Fig. 11. The same as Fig. 10, but for the mass bin centered at 1.03 kg (and ranging between 0.43 and 2.45 kg).

The most critical parameter appears to be the average impact strength Q^* of the targets (i.e., the threshold energy density resulting into catastrophic break-up), which we have assumed to be independent of size and altitude. With $Q^* = 10^4$ and 5×10^4 J/kg, the "catastrophe" in the 900 to 1000 km shell is delayed until about 250 and 400 years in the future, respectively. While recent data (obtained in experiments carried out for military purposes, see Nagl et al. 1992) would suggest that the typical catastrophic break-up threshold is of about 4×10^4 J/kg, fairly close to our highest value, the sensitive dependence of the evolution on Q^* highlights the need for further experimental work to obtain reliable estimates of this parameter for the existing orbiting objects, rather than the potential utility of hardening future satellites.

Concerning future launch policies, we have also tested some fairly "radical" options. Stopping altogether the launching activity 50 years in the future – or, better, assuming that since that time an old satellite is de-orbited for any newly launched one – does not delay the "catastrophe" in any significant way, but just changes the subsequent trends in fragment abundances (which reach a peak and then drop as all the large objects are eliminated by collisional break-up). Even in the scenario of zero net launch rate and no more explosions since the year 2000, the ongoing collisional process will trigger an exponential fragment growth phase and a corresponding rapid decrease of the spacecraft population within the next century. A more realistic option is possibly that of deorbiting half of the satellites after 10 years since their launch.

If this practice were started now (preventing also all the explosions), the catastrophe would be delayed only to about 70 yr in the future; the delay with respect to the nominal case would be even smaller if the deorbiting of old satellites were started only in year 2000. The runaway fragment growth occurs even if, starting 10 yr since now, for every new launch in the three heaviest mass bins (i.e., those involving bodies exceeding about 80 kg), two old satellites of similar mass were deorbited; even in this case the decline in the population of sizeable objects is not fast enough to prevent the fragment population explosion.

Finally, we have tested the possibility of actively removing from space small orbital debris, possibly by using high-power laser sweepers (Schall, 1991, 1993). We have assumed that – besides no more explosions and the two-for-one deorbiting policy described earlier – 1000 objects per year are removed from each of the shells between 700 and 900 km with masses smaller than 13.5 g. Although in this case the fragment populations in the relevant shells are much reduced with respect to the nominal case, the onset of the exponential growth process is not prevented.

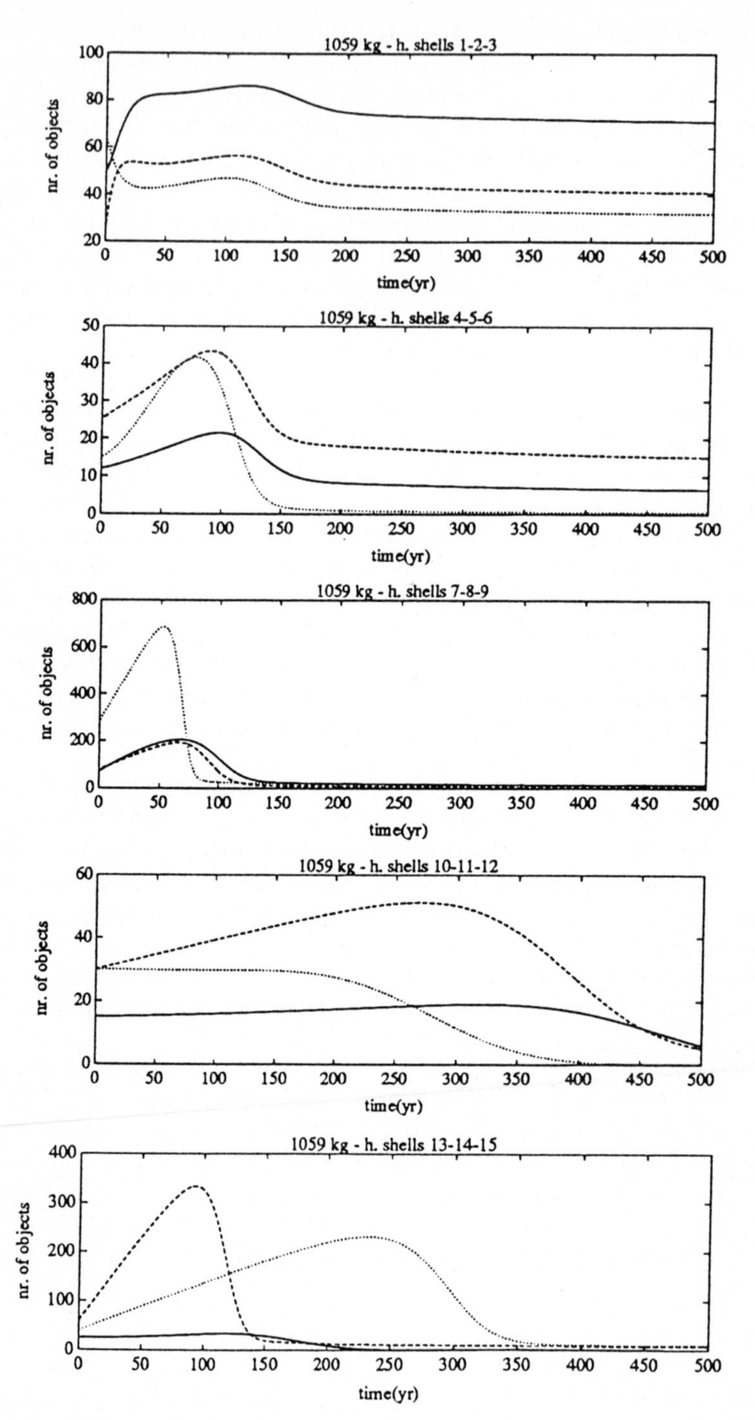

Fig. 12. The same as Figs. 10 and 11, but for the mass bin centered at 1059 kg (and ranging between 445 and 2521 kg).

4 Policy Issues and Conclusions

The space debris problem is a tipical environmental problem. Some new "space" becomes available for human activity. It is clean, wild and economically useful: "a new frontier". Thus people start to exploit, change and pollute it. And after a while we realize that it is too late: the environment is compromised and becomes less and less safe and useful for constructive purposes. This sequence has happened many times over with different niches of the environment on our planet, including the oceans. And although no wildlife is present in circumterrestrial space, its "pollution" by orbital debris will make it soon unavailable for most human activities, unless appropriate measures are urgently taken.

Our modelling work and the corresponding simulations suggest that in some critical altitude shells we have already reached a kind of critical density of objects, which will lead to a catastrophic collisional "chain reaction" in a matter of decades to centuries. No "realistic", stepwise policy option appears suited to prevent such an outcome.

Actually, the need to change the manner in which space activities are conducted has been debated since many years. The options available to decrease the growth of orbital debris depend greatly on the altitude of the mission, the design of the hardware and the commitment of the international spacefaring community. The amount of debris can be controlled in one of two ways: debris prevention and debris removal. An obvious preventive measure is to design missions in order to minimize the lifetime in orbit of the upper stages. Also, in the operation routine of a satellite the covers and all the "operational debris" should be retained. All the residual propellent and pressurant should be expelled from the upper stages left in orbit; this is already being done by the main space agencies since several years (especially for rockets which put payloads in sun-synchronous orbits). Precautions should be taken to prevent break-up due to the overcharging of the batteries. And of course, all the intentional explosions (e.g. for antisatellite weapon testing) should be avoided or, better, banned by appropriate international agreements.

Active removal options include controlled deorbiting, which is feasible and not too expensive at low altitudes, and retrieval of spent or malfunctioning satellites, which has been tried on a few occasions by the US Space Shuttle (but is complex and costly). Of paramount importance in order to lower the break-up probability it would also be the reorbiting in suitable disposal orbits of spent hardware, which at high altitudes is more economical than deorbiting into the Earth's atmosphere. Such a reorbiting routine is particularly important for objects in GEOs, since in this case a population of hundreds of large spacecraft is located in a relatively restricted region of space; for geosynchronous satellites a minimum orbit-raising altitude of 300-400 km is recommended. The use of some types of debris sweeper mechanism (lasers, tethers, etc.) has been widely discussed in recent times, but the corresponding technologies appear not be mature enough yet. Possibly such methods will become feasible (and affordable) in the next century, but for the time being only research and testing in this area can be envisaged.

Some of the above-mentioned measures may be quite expensive and put an additional economical burden on space activities. For instance, in order to de- or re-orbit a spacecraft from the GEO ring some propellant is required that could otherwise be used for station-keeping, thus decreasing the operational lifetime. Thus, a cost to benefit balance argument is made by the launching organizations. Althugh this is unavoidable, like in other areas of environmental policy it entails the risk of making short-sighted choices, which will be regretted in the future.

Also for this reason, a number of steps could be taken in a co-ordinated and legally binding way at an international level, following new agreements and/or treaties to be negotiated at different fora (such as COSPAR, the Geneva UN Conference on Disarmament, the UN Committee for Peaceful Uses of Outer Space). They might be:

- **Creating an internationally available, real-time updated database on the Earth-orbiting population, building upon the 1975 UN Registration Convention and possibly under the responsibility of an *ad hoc* Earth-to-space monitoring Agency.** The rationale for strengthening the Registration Convention and the technical requirements for setting up an international Agency has been discussed in detail in a series of papers by our group (Anselmo et al. 1984, 1991; Bertotti and Farinella 1988; Farinella and Anselmo 1992), and is currently the subject of a specific study by a working group of the UN Institute for Disaramament Research (UNIDIR), whose final report will be published soon. Here it suffices to remark that in a first stage the Agency could just be charged with data archiving and dissemination tasks (with limited, suitably defined confidentiality constraints) and later on it could set up its own data collection, monitoring and verification activities.
- **Extending and making public the available observational and experimental data on satellite break-up events, both occurred in space and simulated in the laboratory.** As we have discussed in Sec. 3, most data on hypervelocity impacts in space are currently classified, and having them freely available would be extremely useful to better assess the debris proliferation hazard.
- **Prohibiting intentional explosions/collisions in space and agreeing on measures aimed at preventing accidental ones, as well as the release of operational debris during launch and orbit injection.** As for deliberate explosions and collisions, it can be noted that the prohibition of such a "polluting" activity would apply regardless from its purpose. But of course such a prohibition would strongly constrain the further development and testing of destabilizing antisatellite weapons. As for accidental explosions, we recall that the main space agencies already have set in place specific guidelines aimed at their prevention (e.g., the venting of unspent propellant and gases from rocket upper stages), so it would just be necessary to extend and standardize the existing rules.
- **Agreeing on a set of international standards to strengthen and/or shield satellites against impact break-up.** Again, such measures are

already being taken by individual space agencies (especially for manned spacecraft), and could well be agreed upon at an international level.

- **Making any launching State/Agency responsible for retrieving or deorbiting satellites and rockets at the end of their mission lifetimes, and removing any other object inserted into orbit within a definite time interval.** This would just be an application to the space environment of the well-known principle that "who causes pollution, is responsible for cleaning up". Also, as we noted earlier moving non-operational satellites into a different orbit, where the collision hazard is less significant, is already common practice in the geostationary ring. Of course, this entails some additional propulsion requirements, and hence costs, on the payloads inserted into orbit; but again in analogy with the terrestrial environment, such additional costs should not be avoided if a sustainable space activity is to be carried out.
- **Analyzing and planning for a future international co-operation in "cleaning-up" the circumterrestrial shells where the debris proliferation hazard is most acute.** This would mean first of all removing from the most proliferation-prone circumterrestrial shells the potential targets for future impacts, i.e. the largest dead satellites and rockets. Optimizing such an activity from the point of view of the cost-to-benefit ratio should be the purpose of specific technical studies to be performed as soon as possible.

Let us end by recalling that in 1878 the first National Park in the world was instituted at Yellowstone *"... for all the people and the whole world, so that it could admire it and enjoy it forever"*. Times appear ripe to act in a similar way for the endangered circumterrestrial outer space.

5 Post-Scriptum

This chapter has been written in 1994. Since then, the space debris issue became even more serious and gained importance inside the space agencies operations.

Although the numbers in Sect. 2, concerning the present orbiting population, have been updated in order to describe the situation at the middle of 1997, the text does not consider some events which have happened in the time span between 1994 and 1997.

First of all, in July 1996, the first recorded accidental collision between an operative spacecraft and a catalogued piece of debris took place: the French microsatellite *Cerise* has been hit by a debris coming from the fragmentation of the 3rd stage of the Ariane V16 rocket (Alby et al. 1997).

Then a new population of particles, composed by more than 14,000 objects larger than 1 cm, has been detected, by radar observations, in the altitude shell between 850 and 1000 km. They have been identified as drops of NaK liquid metal, used as a coolant in the nuclear reactors on board the soviet Radar Ocean Reconnaissance Satellites (RORSATs). These drops leaked outside the radiator pipes when the core was ejected from the reactor, in order to minimize the risk of an uncontrolled reentry of radioactive material inside the Earth's atmosphere.

This set of non fragmentation debris represent, by far, the largest population of cm-sized objects in the largely populated shell around 900 and will increase the number of damaging impacts on operational spacecraft in the next years (Kessler et al. 1997; Rossi et al. 1997).

Other important events, which took place in these last years, are the first launches of satellites composing large constellations of low Earth orbiting spacecraft, mainly devoted to mobile comunications. Many constellations, composed by several dozens of satellites, have been planned and they will represent a big growth in the LEO population (Battaglia and Rossi 1996; Rossi et al. 1997).

References

Alby, F., Lansard, E., Michal, T. (1997): Collision of Cerise with space debris. In *Proceedings of the Second European Conference on Space Debris* (ESOC, Darmstadt, Germany), 589–596

Anselmo, L., Bertotti, B., Farinella, P. (1984): Security in space. In J. Rotblat and A. Pascolini (eds.), *The Arms Race at a Time of Decision* (Macmillan, London), 25–35

Anselmo, L., Bertotti, B., Farinella, P. (1991): International surveillance of outer space for security purposes. Space Policy (August).

Battaglia, P., Rossi, A. (1996): A survey of the italian space debris related activities. In *Proceedings of the* 47th *International Astronautical Congress* (IAA.-96-IAA.6.4.08, Beijing, China, October 7-11)

Bertotti, B., Farinella, P. (1988): Space weapons and arms control. In D. Carlton and C. Schaerf (eds.), *The Arms Race in the Era of Star Wars* (Macmillan, London), 257–271

Bess, T.D. (1975): Mass distribution of orbiting man-made space debris. NASA Tech. Note TDN-8108

Cordelli, A., Farinella, P., Anselmo, L., Pardini, C., Rossi, A. (1993): Future collisional evolution of Earth-orbiting debris. Adv. Space Res. **13**, 8, 215–219

ESA Space Debris Working Group (1988): Space Debris. ESA SP-1109

Farinella, P., Cordelli, A. (1991): The proliferation of orbiting fragments: A simple mathematical model. Science and Global Security **2**, 365–378

Farinella, P., Davis, D.R. (1992): Collision rates and impact velocities in the main asteroid belt. Icarus **97**, 111–123

Fujiwara, A., Kamimoto, G., Tsukamoto, A. (1977): Destruction of basaltic bodies by high-velocity impact. Icarus **31**, 277–288

Fujiwara, A., Cerroni, P., Davis, D.R., Ryan, E., Di Martino, M., Holsapple, K., Housen, K. (1989): Experiments and scaling laws on catastrophic collisions. In R.P. Binzel, T. Gehrels and M.S. Matthews (eds.), *Asteroids II* (Univ. Arizona Press), 240–265

Kessler, D.J., Matney, M.J., Reynolds, R.C., Bernhard, R.P., Stansbery, E.G., Johnson, N.L., Potter, A.E., Anz-Meador, D. (1997): A search for a previously unknown source of orbital debris: the possibility of a coolant leak. In *Radar Ocean Reconnaissance Satellites* (48th International Astronautical Congress, IAA.-97-IAA.6.3.03, Turin, Italy, October 6-10)

Nagl, L., McKnight, D., Maher, R. (1992): Review of data used to support breakup modelling. Paper presented at AIAA/AAS Astrodynamics Conference (Hilton Head Island, South Carolina, August 10-12)

Nebolsine, P.E., et al. (1983): Debris Characterization final report. Prepared by Physical Sciences Inc. for Teledyne Brown Engeneering under prime contract No. G60-84-C-0005

Office of Technology Assessment (1990): *Orbital Debris: A Space Environmental Problem* (US Government Printing Office, Washington DC)

Petit, J.-M., Farinella, P. (1993): Modelling the outcomes of high-velocity impacts between small solar system bodies. Celest. Mech. **57**, 1–28

Rossi, A., Cordelli, A., Farinella, P., Anselmo, L. (1994): Collisional evolution of Earth's orbital debris cloud. J. Geophys. Res. **99**, n^o E11, 23,195–23,210

Rossi, A., Farinella, P. (1992): Collision rates and impact velocities for bodies in low Earth orbit. ESA Journal **16**, 339–348

Rossi, A., Pardini, C., Anselmo, L., Cordelli, A., Farinella, P. (1997): Effects of the RORSAT NaK drops on the long term evolution of the space debris population. In *Radar Ocean Reconnaissance Satellites* (48th International Astronautical Congress, IAA.-97-IAA.6.4.07, Turin, Italy, October 6-10)

Schall, W.O. (1991): Removing small debris from earth orbit. Z. Flugwiss. Weltraumforsch. **15**, 333–341

Schall, W.O. (1993): Active shielding and reduction of the number of small debris with high-power lasers. Presented at First European Conference on Space Debris (Darmstadt, Germany, 5-7 April)

Su, S.-Y., Kessler, D.J. (1985): Contribution of explosions and future collision fragments to the orbital debris environment. Adv. Space Res. **5**, 25–34

Wetherill., G.W. (1967): Collisions in the asteroid belt. J. Geophys. Res. **72**, 2429–2444

Lecture Notes in Physics

For information about Vols. 1–469
please contact your bookseller or Springer-Verlag

Vol. 470: E. Martínez-González, J. L. Sanz (Eds.), The Universe at High-z, Large-Scale Structure and the Cosmic Microwave Background. Proceedings, 1995. VIII, 254 pages. 1996.

Vol. 471: W. Kundt (Ed.), Jets from Stars and Galactic Nuclei. Proceedings, 1995. X, 290 pages. 1996.

Vol. 472: J. Greiner (Ed.), Supersoft X-Ray Sources. Proceedings, 1996. XIII, 350 pages. 1996.

Vol. 473: P. Weingartner, G. Schurz (Eds.), Law and Prediction in the Light of Chaos Research. X, 291 pages. 1996.

Vol. 474: Aa. Sandqvist, P. O. Lindblad (Eds.), Barred Galaxies and Circumnuclear Activity. Proceedings of the Nobel Symposium 98, 1995. XI, 306 pages. 1996.

Vol. 475: J. Klamut, B. W. Veal, B. M. Dabrowski, P. W. Klamut, M. Kazimierski (Eds.), Recent Developments in High Temperature Superconductivity. Proceedings, 1995. XIII, 362 pages. 1996.

Vol. 476: J. Parisi, S. C. Müller, W. Zimmermann (Eds.), Nonlinear Physics of Complex Systems. Current Status and Future Trends. XIII, 388 pages. 1996.

Vol. 477: Z. Petru, J. Przystawa, K. Rapcewicz (Eds.), From Quantum Mechanics to Technology. Proceedings, 1996. IX, 379 pages. 1996.

Vol. 478: G. Sierra, M. A. Martín-Delgado (Eds.), Strongly Correlated Magnetic and Superconducting Systems. Proceedings, 1996. VIII, 323 pages. 1997.

Vol. 479: H. Latal, W. Schweiger (Eds.), Perturbative and Nonperturbative Aspects of Quantum Field Theory. Proceedings, 1996. X, 430 pages. 1997.

Vol. 480: H. Flyvbjerg, J. Hertz, M. H. Jensen, O. G. Mouritsen, K. Sneppen (Eds.), Physics of Biological Systems. From Molecules to Species. X, 364 pages. 1997.

Vol. 481: F. Lenz, H. Grießhammer, D. Stoll (Eds.), Lectures on QCD. VII, 276 pages. 1997.

Vol. 482: X.-W. Pan, D. H. Feng, M. Vallières (Eds.), Contemporary Nuclear Shell Models. Proceedings, 1996. XII, 309 pages. 1997.

Vol. 483: G. Trottet (Ed.), Coronal Physics from Radio and Space Observations. Proceedings, 1996. XVII, 226 pages. 1997.

Vol. 484: L. Schimansky-Geier, T. Pöschel (Eds.), Stochastic Dynamics. XVIII, 386 pages. 1997.

Vol. 485: H. Friedrich, B. Eckhardt (Eds.), Classical, Semiclassical and Quantum Dynamics in Atoms. VIII, 341 pages. 1997.

Vol. 486: G. Chavent, P. C. Sabatier (Eds.), Inverse Problems of Wave Propagation and Diffraction. Proceedings, 1996. XV, 379 pages. 1997.

Vol. 487: E. Meyer-Hofmeister, H. Spruik (Eds.), Accretion Disks – New Aspects. Proceedings, 1996. XIII, 356 pages. 1997.

Vol. 488: B. Apagyi, G. Endrédi, P. Lévay (Eds.), Inverse and Algebraic Quantum Scattering Theory. Proceedings, 1996. XV, 385 pages. 1997.

Vol. 489: G. M. Simnett, C. E. Alissandrakis, L. Vlahos (Eds.), Solar and Heliospheric Plasma Physics. Proceedings, 1996. VIII, 278 pages. 1997.

Vol. 490: P. Kutler, J. Flores, J.-J. Chattot (Eds.), Fifteenth International Conference on Numerical Methods in Fluid Dynamics. Proceedings, 1996. XIV, 653 pages. 1997.

Vol. 491: O. Boratav, A. Eden, A. Erzan (Eds.), Turbulence Modeling and Vortex Dynamics. Proceedings, 1996. XII, 245 pages. 1997.

Vol. 492: M. Rubí, C. Pérez-Vicente (Eds.), Complex Behaviour of Glassy Systems. Proceedings, 1996. IX, 467 pages. 1997.

Vol. 493: P. L. Garrido, J. Marro (Eds.), Fourth Granada Lectures in Computational Physics. XIV, 316 pages. 1997.

Vol. 494: J. W. Clark, M. L. Ristig (Eds.), Theory of Spin Lattices and Lattice Gauge Models. Proceedings, 1996. XI, 194 pages. 1997.

Vol. 495: Y. Kosmann-Schwarzbach, B. Grammaticos, K. M. Tamizhmani (Eds.), Integrability of Nonlinear Systems. XX, 404 pages. 1997.

Vol. 496: F. Lenz, H. Grießhammer, D. Stoll (Eds.), Lectures on QCD. VII, 483 pages. 1997.

Vol. 497: J.P. De Greve, R. Blomme, H. Hensberge (Eds.), Stellar Atmospheres: Theory and Observations. Proceedings, 1996. VIII, 352 pages. 1997.

Vol. 498: Z. Horváth, L. Palla (Eds.), Conformal Field Theories and Integrable Models. Proceedings, 1996. X, 251 pages. 1997.

Vol. 499: K. Jungmann, J. Kowalski, I. Reinhard, F. Träger (Eds.), Atomic Physics Methods in Modern Research. IX, 448 pages. 1997.

Vol. 500: D. Joubert (Ed.), Density Functionals: Theory and Applications. XVI, 194 pages. 1998.

Vol. 501: J. Kertész, I. Kondor (Eds.), Advances in Computer Simulation. VIII, 166 pages. 1998.

Vol. 502: H. Aratyn, T. D. Imbo, W.-Y. Keung, U. Sukhatme (Eds.), Supersymmetry and Integrable Models. XI, 379 pages. 1998.

Vol. 503: J. Parisi, S. C. Müller, W. Zimmermann (Eds.), A Perspective Look at Nonlinear Media. From Physics to Biology and Social Sciences. VIII, 372 pages. 1998.

Vol. 505: D. Benest, C. Froeschlé (Eds.), Impacts on Earth. XVII, 223 pages. 1998.

New Series m: Monographs

Vol. m 2: P. Busch, P. J. Lahti, P. Mittelstaedt, The Quantum Theory of Measurement. XIII, 165 pages. 1991. Second Revised Edition: XIII, 181 pages. 1996.

Vol. m 3: A. Heck, J. M. Perdang (Eds.), Applying Fractals in Astronomy. IX, 210 pages. 1991.

Vol. m 4: R. K. Zeytounian, Mécanique des fluides fondamentale. XV, 615 pages, 1991.

Vol. m 5: R. K. Zeytounian, Meteorological Fluid Dynamics. XI, 346 pages. 1991.

Vol. m 6: N. M. J. Woodhouse, Special Relativity. VIII, 86 pages. 1992.

Vol. m 7: G. Morandi, The Role of Topology in Classical and Quantum Physics. XIII, 239 pages. 1992.

Vol. m 8: D. Funaro, Polynomial Approximation of Differential Equations. X, 305 pages. 1992.

Vol. m 9: M. Namiki, Stochastic Quantization. X, 217 pages. 1992.

Vol. m 10: J. Hoppe, Lectures on Integrable Systems. VII, 111 pages. 1992.

Vol. m 11: A. D. Yaghjian, Relativistic Dynamics of a Charged Sphere. XII, 115 pages. 1992.

Vol. m 12: G. Esposito, Quantum Gravity, Quantum Cosmology and Lorentzian Geometries. Second Corrected and Enlarged Edition. XVIII, 349 pages. 1994.

Vol. m 13: M. Klein, A. Knauf, Classical Planar Scattering by Coulombic Potentials. V, 142 pages. 1992.

Vol. m 14: A. Lerda, Anyons. XI, 138 pages. 1992.

Vol. m 15: N. Peters, B. Rogg (Eds.), Reduced Kinetic Mechanisms for Applications in Combustion Systems. X, 360 pages. 1993.

Vol. m 16: P. Christe, M. Henkel, Introduction to Conformal Invariance and Its Applications to Critical Phenomena. XV, 260 pages. 1993.

Vol. m 17: M. Schoen, Computer Simulation of Condensed Phases in Complex Geometries. X, 136 pages. 1993.

Vol. m 18: H. Carmichael, An Open Systems Approach to Quantum Optics. X, 179 pages. 1993.

Vol. m 19: S. D. Bogan, M. K. Hinders, Interface Effects in Elastic Wave Scattering. XII, 182 pages. 1994.

Vol. m 20: E. Abdalla, M. C. B. Abdalla, D. Dalmazi, A. Zadra, 2D-Gravity in Non-Critical Strings. IX, 319 pages. 1994.

Vol. m 21: G. P. Berman, E. N. Bulgakov, D. D. Holm, Crossover-Time in Quantum Boson and Spin Systems. XI, 268 pages. 1994.

Vol. m 22: M.-O. Hongler, Chaotic and Stochastic Behaviour in Automatic Production Lines. V, 85 pages. 1994.

Vol. m 23: V. S. Viswanath, G. Müller, The Recursion Method. X, 259 pages. 1994.

Vol. m 24: A. Ern, V. Giovangigli, Multicomponent Transport Algorithms. XIV, 427 pages. 1994.

Vol. m 25: A. V. Bogdanov, G. V. Dubrovskiy, M. P. Krutikov, D. V. Kulginov, V. M. Strelchenya, Interaction of Gases with Surfaces. XIV, 132 pages. 1995.

Vol. m 26: M. Dineykhan, G. V. Efimov, G. Ganbold, S. N. Nedelko, Oscillator Representation in Quantum Physics. IX, 279 pages. 1995.

Vol. m 27: J. T. Ottesen, Infinite Dimensional Groups and Algebras in Quantum Physics. IX, 218 pages. 1995.

Vol. m 28: O. Piguet, S. P. Sorella, Algebraic Renormalization. IX, 134 pages. 1995.

Vol. m 29: C. Bendjaballah, Introduction to Photon Communication. VII, 193 pages. 1995.

Vol. m 30: A. J. Greer, W. J. Kossler, Low Magnetic Fields in Anisotropic Superconductors. VII, 161 pages. 1995.

Vol. m 31: P. Busch, M. Grabowski, P. J. Lahti, Operational Quantum Physics. XI, 230 pages. 1995.

Vol. m 32: L. de Broglie, Diverses questions de mécanique et de thermodynamique classiques et relativistes. XII, 198 pages. 1995.

Vol. m 33: R. Alkofer, H. Reinhardt, Chiral Quark Dynamics. VIII, 115 pages. 1995.

Vol. m 34: R. Jost, Das Märchen vom Elfenbeinernen Turm. VIII, 286 pages. 1995.

Vol. m 35: E. Elizalde, Ten Physical Applications of Spectral Zeta Functions. XIV, 228 pages. 1995.

Vol. m 36: G. Dunne, Self-Dual Chern-Simons Theories. X, 217 pages. 1995.

Vol. m 37: S. Childress, A.D. Gilbert, Stretch, Twist, Fold: The Fast Dynamo. XI, 410 pages. 1995.

Vol. m 38: J. González, M. A. Martín-Delgado, G. Sierra, A. H. Vozmediano, Quantum Electron Liquids and High-T_c Superconductivity. X, 299 pages. 1995.

Vol. m 39: L. Pittner, Algebraic Foundations of Non-Commutative Differential Geometry and Quantum Groups. XII, 469 pages. 1996.

Vol. m 40: H.-J. Borchers, Translation Group and Particle Representations in Quantum Field Theory. VII, 131 pages. 1996.

Vol. m 41: B. K. Chakrabarti, A. Dutta, P. Sen, Quantum Ising Phases and Transitions in Transverse Ising Models. X, 204 pages. 1996.

Vol. m 42: P. Bouwknegt, J. McCarthy, K. Pilch, The $\mathcal{W}_3$ Algebra. Modules, Semi-infinite Cohomology and BV Algebras. XI, 204 pages. 1996.

Vol. m 43: M. Schottenloher, A Mathematical Introduction to Conformal Field Theory. VIII, 142 pages. 1997.

Vol. m 44: A. Bach, Indistinguishable Classical Particles. VIII, 157 pages. 1997.

Vol. m 45: M. Ferrari, V. T. Granik, A. Imam, J. C. Nadeau (Eds.), Advances in Doublet Mechanics. XVI, 214 pages. 1997.

Vol. m 46: M. Camenzind, Les noyaux actifs de galaxies. XVIII, 218 pages. 1997.

Vol. m 47: L. M. Zubov, Nonlinear Theory of Dislocations and Disclinations in Elastic Body. VI, 205 pages. 1997.

Vol. m 48: P. Kopietz, Bosonization of Interacting Fermions in Arbitrary Dimensions. XII, 259 pages. 1997.

Vol. m 49: M. Zak, J. B. Zbilut, R. E. Meyers, From Instability to Intelligence. Complexity and Predictability in Nonlinear Dynamics. XIV, 552 pages. 1997.

Vol. m 50: J. Ambjørn, M. Carfora, A. Marzuoli, The Geometry of Dynamical Triangulations. VI, 197 pages. 1997.

Vol. m 51: G. Landi, An Introduction to Noncommutative Spaces and Their Geometries. XI, 200 pages. 1997.

Vol. m 52: M. Hénon, Generating Families in the Restricted Three-Body Problem. XI, 278 pages. 1997.

Vol. m53: M. Gad-el-Hak, A. Pollard, J.-P. Bonnet (Eds.), Flow Control. Fundamentals and Practices. XII, 527 pages. 1998.